Textbook of
PLANT PHYSIOLOGY

The Authors

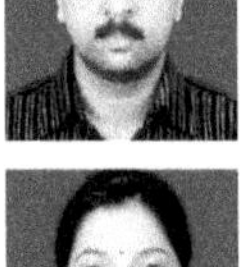

Dr. Narendra Shankar Pandey obtained his M.Sc. (Botany) and Ph.D. from Dr. RML Awadh University, Faizabad. He has published various research appers in national and international journals of repute. He got several prizes and honour. He is an environmentalist and writer.

Dr. Pramila Pandey obtained her Ph.D. (Botany) from Dr. RML Awadh University, Faizabad. She qualified NET and did P.G. Diploma in Bioinformatics. She has published various research papers in national and international journals of repute. Besides this, she has participated/delivered lectures at various conferences/sympisium *etc.* Presently Dr. Pandey is working as Programme Assistant in the Department of Biotechnology at N.D. University of Agriculture and Technology, Kumarganj, Faizabad.

Textbook of
PLANT PHYSIOLOGY

Dr. Narendra Shankar Pandey
Dr. Pramila Pandey

2016
Daya Publishing House®
A Division of
Astral International Pvt. Ltd.
New Delhi – 110 002

Cataloging in Publication Data--DK
Courtesy: D.K. Agencies (P) Ltd. <docinfo@dkagencies.com>

Pandey, Narendra Shankar, author.
Textbook of plant physiology / Dr. Narendra Shankar Pandey, Dr. Pramila Pandey.
pages cm
Includes index.

ISBN 978-93-5130-990-1 (International Edition)

1. Plant physiology--Textbooks. I. Pandey, Pramila, author. II. Title. III. Title: Plant physiology.

QK711.2.P36 2016 DDC 571.2 23

Published by : **Daya Publishing House®**
A Division of
Astral International Pvt. Ltd.
– ISO 9001:2008 Certified Company –
4760-61/23, Ansari Road, Darya Ganj
New Delhi-110 002
Ph. 011-43549197, 23278134
E-mail: info@astralint.com
Website: www.astralint.com

Laser Typesetting : **Classic Computer Services**, Delhi - 110 035

Printed at : **Sanat Printers**

Preface

We feel delighted to write the preface for the "Textbook of Plant Physiology". Yet as plant biologists we realize that time always limits what can be presented in a plant physiology course. Hence, it has been our assignment to choose and summarize those new finding that are most representative of the status of the science. We hope that students of undergraduate level and postgraduate as well as for those student who are preparing for competitive exam will agree with many of our choices. While writing the book we keep in the mind the syllabus of all universities and competitive exams. Special attempt has been made to make the book useful to students preparing for competitive examinations.

We are grateful to all those editors, authors and publishers whose work we have consulted in writing this book. We have no words of gratitude for my friends and students who helped me a lot and inspired all along during the preparation of this book.

We would like to express my thanks to also my publisher Astral International Pvt. Ltd. for kindly publishing this book.

The authors will greatly appreciate knowing reader's criticism and suggestions for the improvement of this book will be thankfully acknowledge on authors e-mail pramila28@gmail.com and narendramilkyway@gmail.com.

Narendra Shankar Pandey

Pramila Pandey

Contents

Chapter 1

Water Metabolism

1.1. The Structure and Properties of Water

The water molecule consists of an oxygen atom covalently bonded to two hydrogen atoms. The two O–H bonds form an angle of 105° (Figure 1.1). Because the oxygen atom is more electronegative than hydrogen, it tends to attract the electrons of the covalent bond. This attraction results in a partial negative charge at the oxygen end of the molecule and a partial positive charge at each hydrogen.This asymmetric electron distribution makes water a polar molecule.The strong mutual attraction between adjacent water molecules or between water and other polar molecules. This attraction is called hydrogen bonding.The energy of H-bonding is about 20 kj mol^{-1}. H-bonding is responsible for many unique properties of water.

1. Water is an excellent solvent: It dissolves greater amounts of a wider variety of substances than do other related solvents.

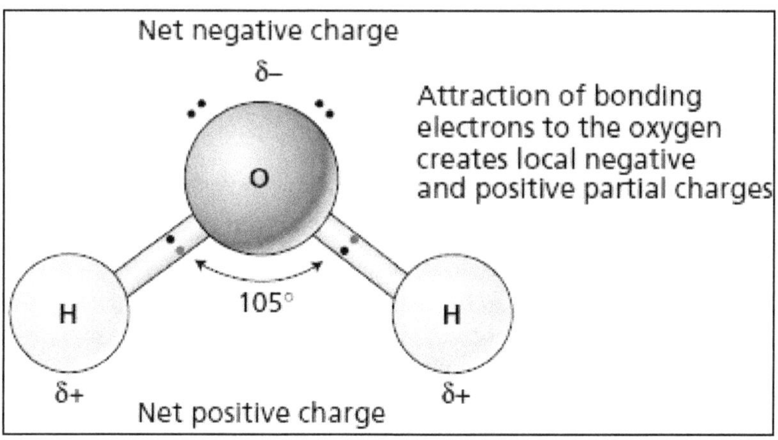

Figure 1.1: Chemical Structure of Water.

2. The extensive hydrogen bonding between water molecules results in unusual thermal properties, such as high specific heat and high latent heat of vaporization.
3. The cohesive and adhesive properties of water are due to hydrogen bonding.
4. Cohesion gives water a high tensile strength, defined as the maximum force per unit area that a continuous column of water can withstand before breaking.

1.2. Imbibition

It is the phenomeno of adsorption of water or any other liquid by the solid particles of a substance without forming a solution. If dry piece of wood is placed in water, it swells and increase in its volume. Similarly, if dry gum or pieces of agar agar are placed in water, they swell and their volume increases. Imbibition is the initial step in the germination of the seed.

1.3. Diffusion

It is a directed movement from a region of a high concentration to a region of lower concentration.Diffusion is,therefore,the movement of molecules or ions of a solute or a solvent, be it a solid,liquid or gas from the region of its higher concentration to that of its lower concentration.

The diffusing molecule or ions exert a pressure called diffusion pressure (DP). The diffusing pressure is directly proportional to the concentration or the number of diffusing particle.

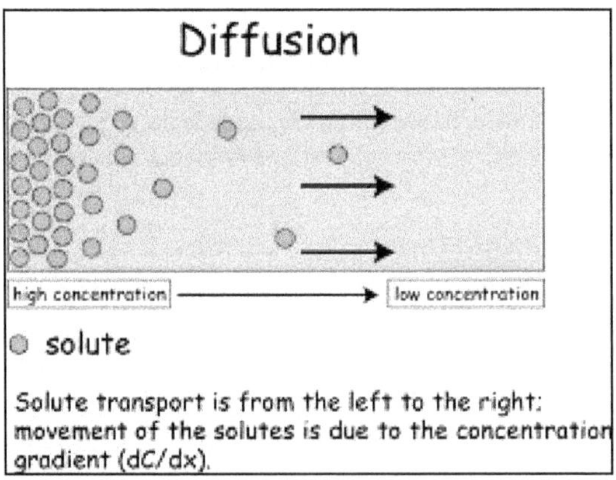

Figure 1.2: Diffusion.

In the 1880s the German scientist Adolf Fick discovered that the rate of diffusion is directly proportional to the concentration gradient ($\Delta cs/\Delta x$)–that is, to the difference in concentration of substance s (Δcs) between two points separated by the distance Δx. In symbols, we write this relation as Fick's first law:

$$\Delta J_s = D_s \frac{\Delta C_s}{\Delta X}$$

1.4. Permeability

A body is said to be permeable to a substance, if it allows the passage of the substance through it.This property is termed as permeability.The biological membrane are not perfectly semipermeable. They allow to pass solute molecules only upto certain extent and and are therefore, selective in nature.They are known as selectively or differentially permeable membranes.

1.5. Osmosis

Osmosis is a type of diffusion. Osmosis can occur when there is a partially permeable membrane, such as a cell membrane. When a cell is submerged in water, the water molecules pass through the cell membrane from an area of low solute concentration to high solute concentration (*e.g.* if the cell is submerged in saltwater, water molecules move out; if it is submerged in freshwater, however, water molecules move in); this is called osmosis. The cell membrane is selectively permeable, so only necessary materials are let into the cell and wastes are left out. The word 'osmosis' is particular to the diffusion of water molecules into the cell.

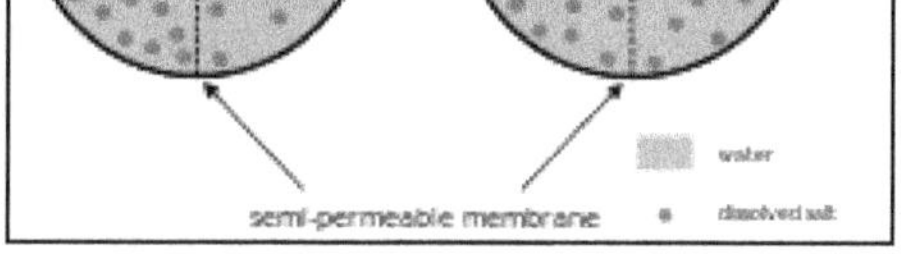

Figure 1.3: Water Passing through a Semi-Permeable Membrane.

When the membrane has a volume of pure water on both sides, water molecules pass in and out in each direction at exactly the same rate; there is no net flow of water through the membrane.

Osmosis can be explained using the concept of thermodynamic free energy: the less concentrated solution contains more free energy, so its solvent molecules will tend to diffuse to a place of lower free energy in order to equalize free energy. Since the semipermeable membrane only allows solvent molecules to pass through it, the result is a net flow of water to the side with the more concentrated solution. Assuming the membrane does not break, this net flow will slow and finally stop as the pressure on the more concentrated side lessens and the movement in each direction becomes equal: this state is called dynamic equilibrium.

When plant cell is placed in pure distilled water,the water enter in to the cell.The plan cell is permeable to both solute and solvent molecule.The plasma membrane,on the oher hand,is differentially permeable.The water molecule enter in to the cell sap as a result of simple process of osmosis. This is known as endosmosis.If the direction of movement is reverse *i.e.,* from cell sap to the outside,the osmosis is termed exosmosis.

When a plant cell is placed in a hypertonic solution, the water in the cells moves to an area higher in solute concentration and the cell shrinks, and in doing so, becomes flaccid. This means the cell has become plasmolyzed – the cell membrane has completely left the cell wall due to lack of water pressure on it; the opposite of turgid.

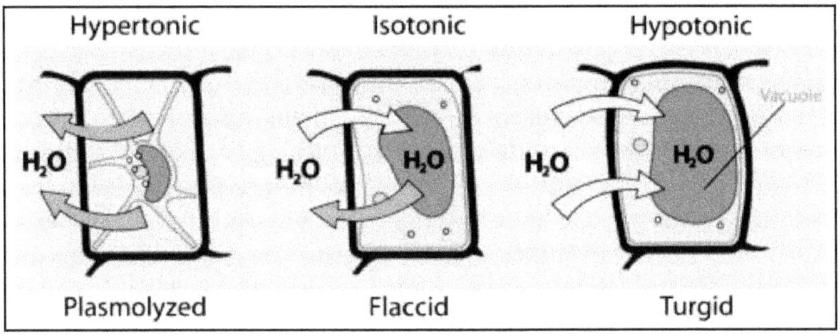

Figure 1.4: Difference between Hypertonic, Hypotonic and Isotonic Solution.

Differences between Diffusion and Osmosis

Diffussion	*Osmosis*
1. Diffussion is the net downward movement of a given substance (ions, atoms, or molecules etc.) along the free energy gradient from the place of its higher concentration to an area of its lesser concentration.	1. Osmosis is a special type of diffusion of solvent molecules from low concentration of solution to higher concentration of solution, when the two are seprated by a semipermeable, membrane.
2. The diffusion may occure in any medium and the diffusing particles may be solid, liquid or gas.	2. The osmosis occurs in liquid medium and only the solvent molecules move from one place to another.
3. Prescence of semipermeable membrane is not required. Diffussion may occure through a membrane where the movement is possible.	3. Prescence of semi-permeable membrane in between the two solutions is required.

1.6. Osmotic Pressure (OP)

Osmotic pressure of a solution is equivalent to the pressure which must be applied to it in order to prevent the flow of solvent due to osmosis. The term osmotic pressure is misleading. The term solute potential (or osmotic potential) is preferred over the osmotic pressure. The solute potential is equal to osmotic pressure. The solute potential is equal to osmotic pressure is magnitude but opposite in sign.

$$\text{Osmotic pressure} = -\text{Solute potential } (-\Psi_s)$$

1.7. Turgor Pressure (TP)

When the water entry in to the cell sap a pressure is developed in the protoplasm which presses against the cell wall.The actual pressure exerted by the protoplasm against the cell wall is the turgor pressure(TP), Cell wallis rigid and elastic, therefore, it exerts an equal and opposite pressure aganist the expanding protoplasm. This pressure is called wall pressure (WP).

$$TP = WP$$

1.8. Diffussion Pressure deficit (DPD) or Suction Pressure

The pressure exerted by the diffusing particles is known as diffusion pressure (DP). The pure water has maximum diffusion pressure.Diffussion pressure of a solvent in a solution is always lower than that of its pure solvent.The amount by which the diffusion pressure of a solution is lower than that of its pure solvent is known as diffusion pressure deficit (DPD).

The relationship of TP, OP,WP and DPD is, therefore as follows,

DPD = OP - WP

But, WP = TP

Therefore, DPD = OP - TP

1.9. Concept of Water Potential

All living things, including plants, require a continuous input of free energy to maintain and repair their highly organized structures, as well as to grow and reproduce. Processes such as biochemical reactions, solute accumulation, and long-distance transport are all driven by an input of free energy into the plant.

The difference between the free energy(free energy is the sum of an energy of a system capable of doing work) of water molecule in pure water and the energy of water in any other system(*e.g.*- water in a solution or in a plant cell or tissue) is termed the water potential. The greek letter psi, the symbol for which is Ψ designates the water potential in a system. The water potential is measured in bars, a bar is a pressure unit.Water potential of protoplasm is equal but opposite in sign to the diffusion pressure deficit (DPD). The water potential of pure water is zero. The presence of solute paricles reduces the free energy of water and thus decreses the water potential (negative value).It is, therefore, the water potential of a solution is always less than zero. The direction of flow will be energetically down hill, that is, from the region of higher water potential to lower water potential. The movement of water continues until the water potential of the two regions become equal.

1.10. The Component of Water Potential

The major factors influencing the water potential in plants are *concentration*, *pressure*, and *gravity*. Water potential is symbolized by Ψ_w (the Greek letter psi), and the water potential of solutions may be dissected into individual components, usually written as the following sum:

$$\Psi_w = \Psi_s + \Psi_p + \Psi_g$$

The terms Ψ_s, Ψ_p, and Ψ_g denote the effects of solutes, pressure, and gravity, respectively, on the free energy of water.

Solutes

The term Ψ_s, called the solute potential or the osmotic potential, represents the effect of dissolved solutes on water potential. Solutes reduce the free energy of water by diluting the water. This is primarily an entropy effect; that is, the mixing of solutes and water increases the disorder of the system and thereby lowers free energy. This means that the osmotic potential is independent of the specific nature of the solute. For dilute solutions of nondissociating substances.

Pressure

The term Ψ_p is the hydrostatic pressure of the solution. Positive pressures raise the water potential; negative pressures reduce it. Sometimes Ψ_p is called pressure potential. The positive hydrostatic pressure within cells is the pressure referred to as turgor pressure. The value of Ψ_p can also be negative, as is the case in the xylem and in the walls between cells, where a tension, or negative hydrostatic pressure, can develop. The negative pressures of outside cells are very important in moving water long distances through the plant.

Gravity

Gravity causes water to move downward unless the force of gravity is opposed by an equal and opposite force. The term Ψ_g depends on the height (*h*) of the water above the reference-state water, the density of water (*r*w), and the acceleration due to gravity (*g*).

When dealing with water transport at the cell level, the gravitational component (Ψ_g) is generally omitted because it is negligible compared to the osmotic potential and the hydrostatic pressure. Thus, in these cases Equation can be simplified as follows:

$$\Psi_w = \Psi_s + \Psi_p$$

The osmotic behavior of plant cells with some numerical examples. First imagine an open beaker full of pure water at 20°C. Because the water is open to the atmosphere, the hydrostatic pressure of the water is the same as atmospheric pressure (Yp = 0 MPa). There are no solutes in the water, so Ys = 0 MPa; therefore the water potential is 0 MPa (Yw = Ys + Yp).

1.11. Water in the Soil

The water content and the rate of water movement in soils depend to a large extent on soil type and soil structure. Table 1.1 shows that the physical characteristics of different soils can vary greatly. At one extreme is sand, in which the soil particles may be 1 mm or more in diameter. Sandy soils have a relatively low surface area per gram of soil and have large spaces or channels between particles. When a soil is heavily watered by rain or by irrigation, the water percolates downward by gravity through the spaces between soil particles, partly displacing, and in some cases trapping, air in these channels. Water in the soil may exist as a film adhering

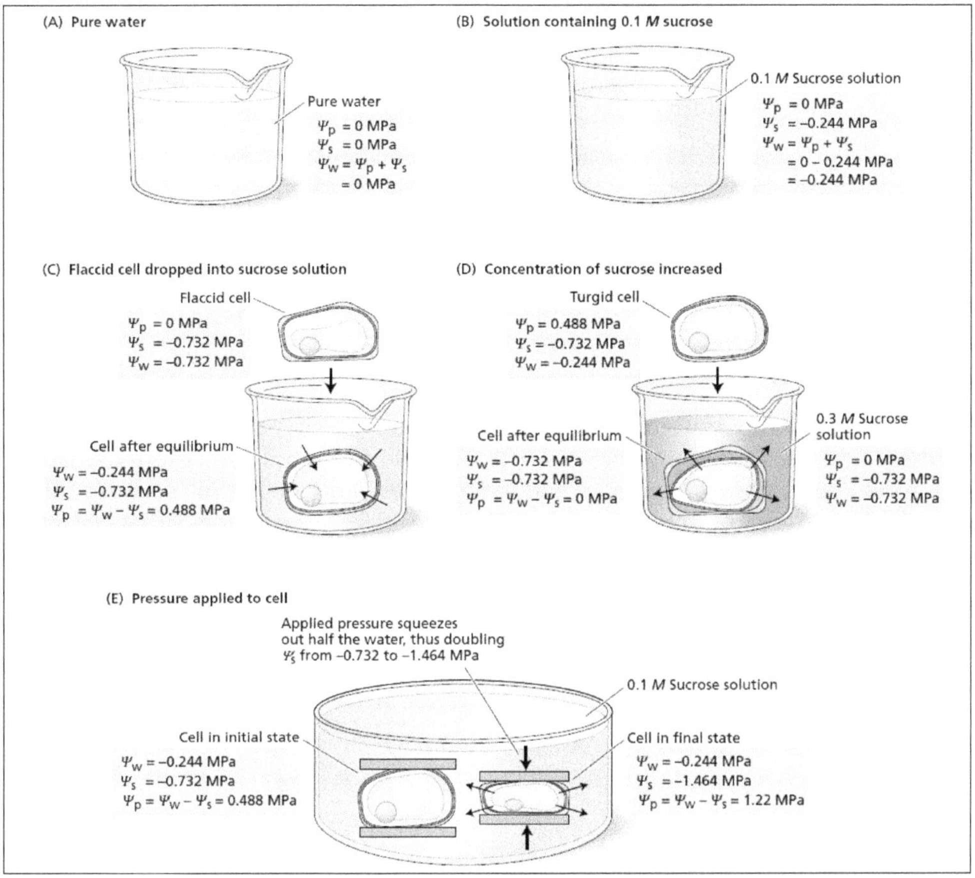

Figure 1.5: Five Examples Illustrating the Concept of Water Potential and its Components.

(A) Pure water. (B) A solution containing 0.1 M sucrose. (C) A flaccid cell (in air) is dropped in the 0.1 M sucrose solution. Because the starting water potential of the cell is less than the water potential of the solution, the cell takes up water. After equilibration, the water potential of the cell rises to equal the water potential of the solution, and the result is a cell with a positive turgor pressure. (D) Increasing the concentration of sucrose in the solution makes the cell lose water. The increased sucrose concentration lowers the solution water potential, draws water out from the cell, and thereby reduces the cell's turgor pressure. In this case the protoplast is able to pull away from the cell wall (i.e, the cell plasmolyzes) because sucrose molecules are able to pass through the relatively large pores of the cell walls. In contrast, when a cell desiccates in air (*e.g.*, the flaccid cell in panel C) plasmolysis does not occur because the water held by capillary forces in the cell walls prevents air from infiltrating into any void between the plasma membrane and the cell wall. (E) Another way to make the cell lose water is to press it slowly between two plates. In this case, half of the cell water is removed, so cell osmotic potential increases by a factor of 2 (Courtesy of Taiz and Zieger).

to the surface of soil particles, or it may fill the entire channel between particles. The moisture-holding capacity of soils is called the field capacity. Field capacity is the water content of a soil after it has been saturated with water and excess water has been allowed to drain away.

Table 1.1: Physical Characteristics of different Soils

Soil	*Particle Diameter (μm)*	*Surace Area per gram (m^2)*
Coarse sand	2000-200	<1-10 (Coarse sand and Fine sand)
Fine sand	200-20	
Silt	20-2	10-100
Clay	<2	100-1000

Like the water potential of plant cells, the water potential of soils may be dissected into two components, the osmotic potential and the hydrostatic pressure. The osmotic potential of soil water is generally negligible because solute concentrations are low; a typical value might be –0.02 MPa. For soils that contain a substantial concentration of salts, however, *Y*s is significant, perhaps –0.2 MPa or lower. The second component of soil water potential is hydrostatic pressure (*Y*p). For wet soils, *Y*p is very close to zero. As a soil dries out, *Y*p decreases and can become quite negative. The capillary water is readily available to the plants and it is the main source.

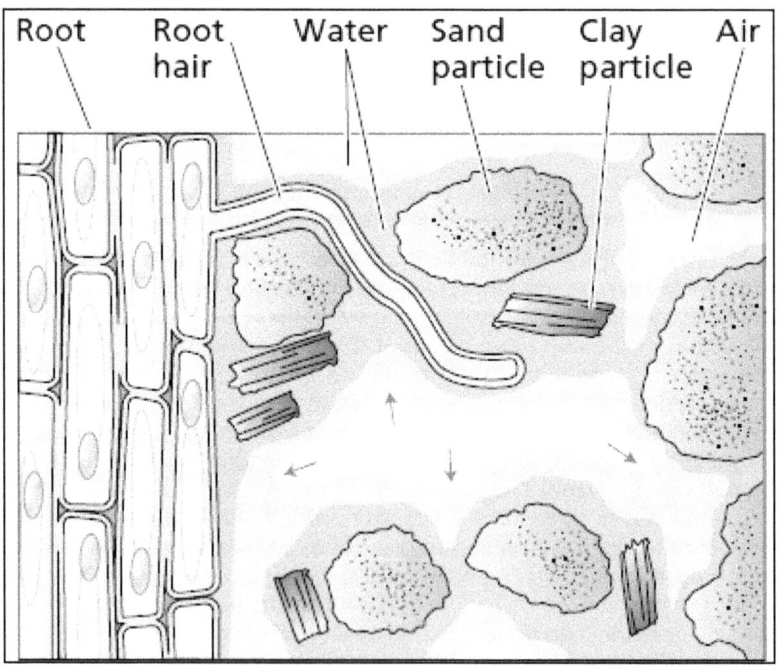

Figure 1.6: Root Make a Intimate Contact with Soil Particle.

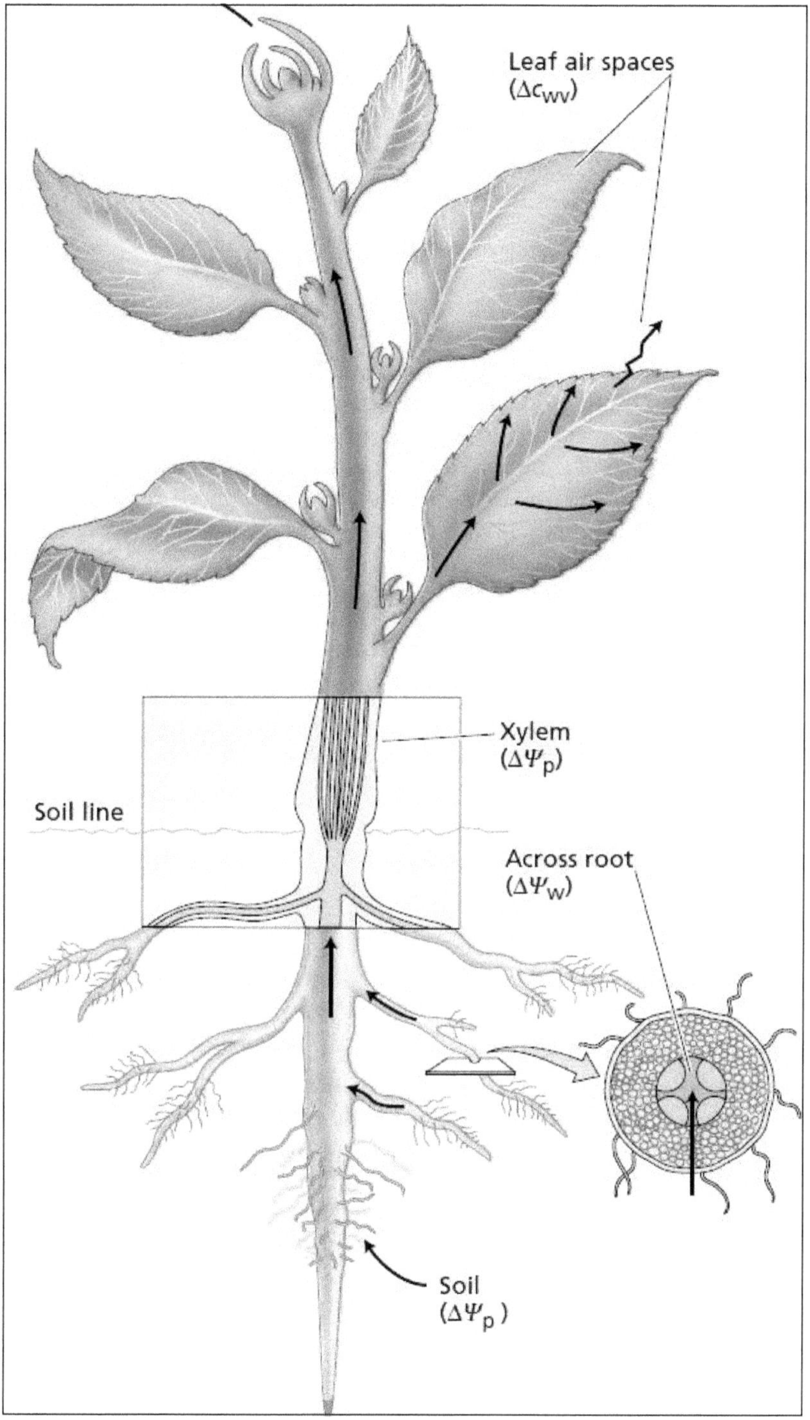

Figure 1.7: Water Flow from Soil through Plant to Atmosphere.

Water moves through soils predominantly by bulk flow driven by a pressure gradient. In addition, diffusion of water vapor accounts for some water movement. As plants absorb water from the soil, they deplete the soil of water near the surface of the roots. This depletion reduces Yp in the water near the root surface and establishes a pressure gradient with respect to neighboring regions of soil that have higher Yp values. Because the water-filled pore spaces in the soil are interconnected, water moves to the root surface by bulk flow through these channels down the pressure gradient.

1.12. Water Absorption

Intimate contact between the surface of the root and the soil is essential for effective water absorption by the root. This contact provides the surface area needed for water uptake and is maximized by the growth of the root and of root hairs into the soil. Root hairs are microscopic extensions of root epidermal cells that greatly increase the surface area of the root, thus providing greater capacity for absorption of ions and water from the soil. Water enters the root most readily in the apical part of the root that includes the root hair zone. More mature regions of the root often have an outer layer of protective tissue, called an *exodermis* or *hypodermis*, that contains hydrophobic materials in its walls and is relatively impermeable to water.

1.12.1. Mechnism of Water Absorption

The initial step in the water uptake is the imbibitions of water molecules by outer pectic layer of the cell wall which is hydrophilic in nature. There are two distinct mechanism which independently operate in the absorption of waer in plants. These are-

1. Active absorption
2. Passive absorption

Active Absorption

When the water is absorbed due to activity of roots itself, particularly root hairs.The active absorption may be osmotic or non-osmotic.

The first step in the absorption of water is its imbibitions on the cell wall surface. The water enters through the cell wall and comes in contact with differential permeable plasma membrane.The osmotic potential of soil water remains below the cell sap. Thus, there exists a great difference in the osmotic potential of two sides. The water molecules moves from the side of higher water potential towards the side of lower water potential.The soil solution, having less osmotic potential (or solute potential), has greater water potential.On the other hand,the cellsap with more OP has more negative water potential.Thus,the water enters into the cellsap due to water potential gradient.Water moves from cell to cell along the concentration gradient and finally reaches up to endodermis.

Passive Absorption

Under condition of passive absorption the factors originating in the upper part of the plant play an important role. Passive water absorption is regulated by transpiration.As the water is lost from the leaf cells their volume falls.Their turgor pressure drops rapidaly and consequently their DPD is increased. Thus they ultimately withdraw water from the adjacent xylem cells.The tension developed at the upper ed of xylem is transmitted downwards and result withdrawal of water from adjoining cells of roots in the root hair zone.the transport of water mainly occurs through the cell walls and intercellular spaces (apoplast).

Difference between Active and Passive Water Absorption

Active Absorption

1. It occurs due to the activity of roots and particularly root hairs
2. It is due to osmotic and non osmotic mechanisms
3. In active absorption the osmotic process involves diffusion pressure deficit(DPD) of the cells. The root hairs have more DPD as compared to soil solution and therefore, water is taken in. Water is still absorbed if soil solution has greater DPD, But under expenditure of metabolic energy.
4. Active absorption involves symplast movement of water in root hairs. The water first enters the cell sap and then passes from one cell to another. Such types of movement where type of movement where living protoplasm involved, is called symplast.
5. Evidences in support of active water absorption are root pressure, bleeding and guttation.

Passive Absorption

1. The passive absorption occurs due to the activity of upper part of plant, such as shoot and leaves.
2. Passive absorption is due to the process of active transpiration in the upper part.
3. The passive absorption occurs due to the tension created in the xylem sap by transpiration pull.
4. In passive absorption water moves probably through the free spaces or apoplast of root. The apoplast path of water movement includes cell wall and intercellular spaces. Which are fully permeable. The water can reach up to endodermis through apoplast but it moves through the endodermis by symplast.
5. Evidences in support of passive water absorption can be given by cutting the roots under water. The absorption of water still occurs if all the roots are removed.

1.13. Water Movement in Root

From the epidermis to the endodermis of the root, there are three pathways through which water can flow (Figure 1.8): the apoplast, transmembrane, and symplast pathways. The concept and terminology of apoplast and symplast were introduced by Munch (1930).

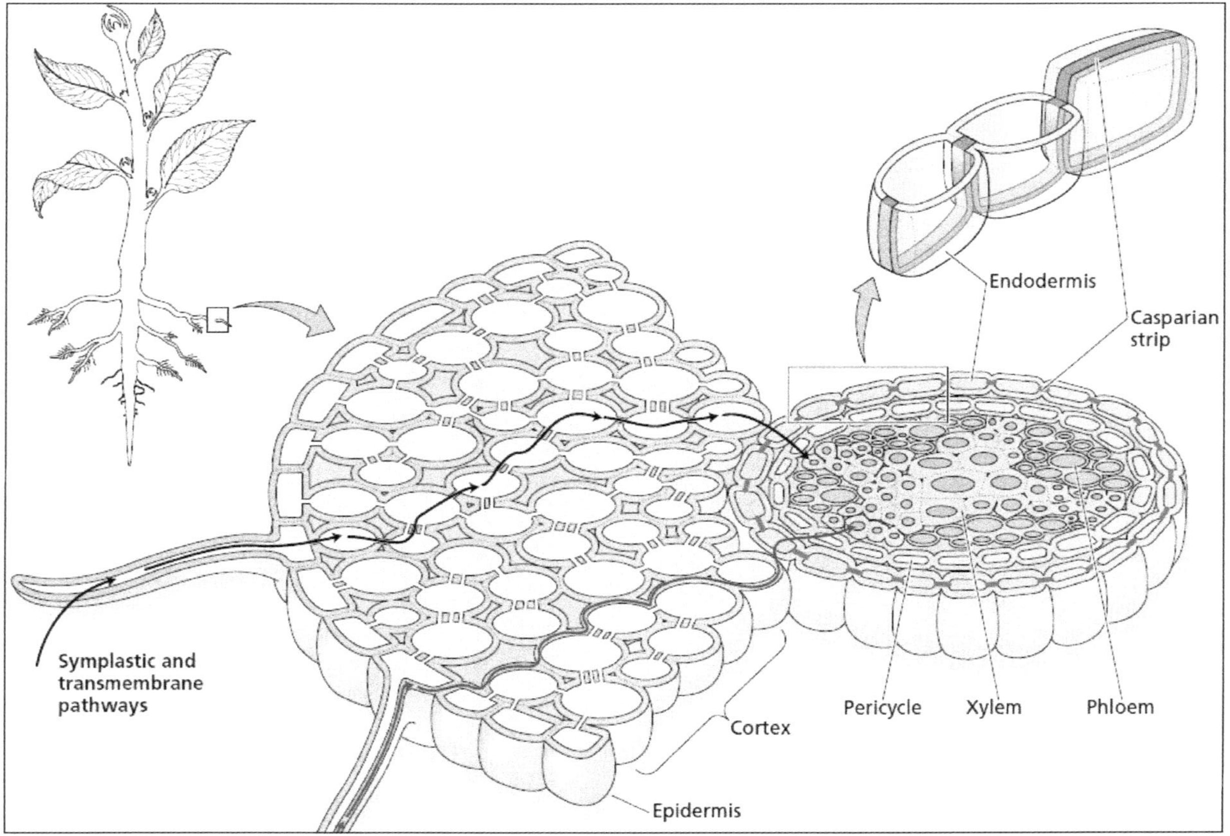

Figure 1.8: Pathways for Water Uptake by the Root. Through the cortex, water may travel via the apoplast pathway, the transmembrane pathway, and the symplast pathway. In the symplast pathway, water flows between cells through the plasmodesmata without crossing the plasma membrane. In the transmembrane pathway, water moves across the plasma membranes, with a short visit to the cell wall space. At the endodermis, the apoplast pathway is blocked by the Casparian strip.

1. In the apoplast pathway, water moves exclusively through the cell wall without crossing any membranes. The apoplast is the continuous system of cell walls and intercellular air spaces in plant tissues.
2. The transmembrane pathway is the route followed by water that sequentially enters a cell on one side, exits the cell on the other side, enters the next in the series, and so on. In this pathway, water crosses at least two membranes for each cell in its path (the plasma membrane on entering and on exiting). Transport across the tonoplast may also be involved.
3. In the symplast pathway, water travels from one cell to the next via the plasmodesmata. The symplast consists of the entire network of cell cytoplasm interconnected by plasmodesmata.

At the endodermis, water movement through the apoplast pathway is obstructed by the Casparian strip (see Figure 1.8). The Casparian strip is a band of radial cell walls in the endodermis that is impregnated with the waxlike, hydrophobic substance suberin. Suberin acts as a barrier to water and solute movement.The Casparian strip breaks the continuity of the apoplast pathway,and forces water and solutes to cross the endodermis by passing through the plasma membrane. Thus, despite the importance of the apoplast pathway in the root cortex and the stele, water movement across the endodermis occurs through the symplast.

Factors Affecting the Rate of Water Absorption

Most of the factors which affect the water absorption are:

1. Available soil water
2. Temperature of the soil – at low temperature of soil (*i.e.* less than 20°C) reduces absorption of water.
3. Concentration of soil solution
4. Aeration of soil –poorly aerated soil,particularly when it is deficient in oxygen,retards absorption of water.Accumulation of CO_2 in the soil also inhibits absorption.

1.14. Ascent of Water (or Translocation)

The water is absorbed mainly by root and is moved upward to all the parts of plants via stem. Thus, the upward movement of water through stem is called ascent of sap.Once the water enters in to the xylem it continues its upward movement until it reaches the mesophyll tissues of the leaves.

Many theories have been put forward to explain the upward movement of water. These are:

1. Root pressure theory
2. Vital theories
3. Physical theories

1.15. Root Pressure Theory

Plants sometimes exhibit a phenomenon referred to as root pressure. For example, if the stem of a young seedling is cut off just above the soil, the stump will often exude sap from the cut xylem for many hours. If a manometer is sealed over the stump, positive pressures can be measured. These pressures can be as high as 0.05 to 0.5 MPa. Roots generate positive hydrostatic pressure by absorbing ions from the dilute soil solution and transporting them into the xylem. The buildup of solutes in the xylem sap leads to a decrease in the xylem osmotic potential (Ψs) and thus a decrease in the xylem water potential (Ψw). This lowering of the xylem Ψ_w provides a driving force for water absorption, which in turn leads to a positive hydrostatic pressure in the xylem. In effect, the whole root acts like an osmotic cell; the multicellular root tissue behaves as an osmotic membrane does, building up a positive hydrostatic pressure in the xylem in response to the accumulation of solutes. Root pressure is most likely to occur when soil water potentials are high and transpiration rates are low. When transpiration rates are high, water is taken up so rapidly into the leaves and lost to the atmosphere that a positive pressure never develops in the xylem. Plants that develop root pressure frequently produce liquid droplets on the edges of their leaves, a phenomenon known as guttation. Positive xylem pressure causes exudation of xylem sap through specialized pores called *hydathodes* that are associated with vein endings at the leaf margin. The "dewdrops" that can be seen on the tips of grass leaves in the morning are actually guttation.

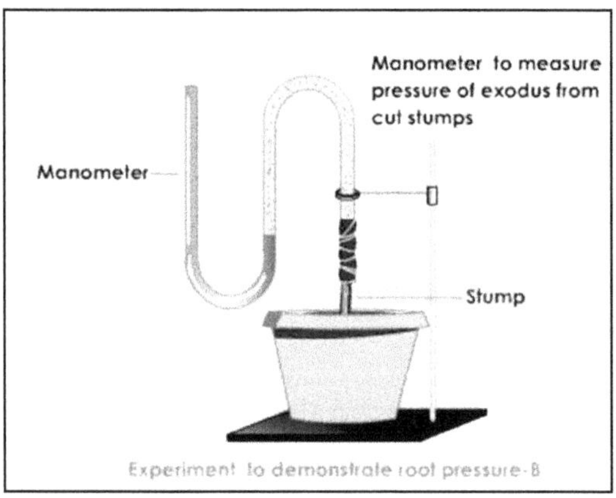

Figure 1.9: Experiment to Demonstrate Root Pressure.

1.16. Vital Theory

"Vital force theory" is a proposed mechanism for the ascent of sap through the xylem tissue of plants.

According to the vital force theory, the conduction of water up the xylem vessel is a result of vital action of the living cells in the xylem tissue. Jagadish Chandra Bose suggested a mechanism for the ascent of sap in 1927. He found electrical 'pulsations' or oscillations in electric potentials, and came to believe these were coupled with rhythmic movements in the telegraph plant Codariocalyx motorius (then Desmodium). On the basis of this Bose theorized that regular wave-like 'pulsations' in cell electric potential and turgor pressure were an endogenous form of cell signaling. According to him the living cells in the inner lining of the xylem tissue pump water by contractive and expulsive movements similar to the animal heart circulating blood. This mechanism has not been well supported, and in spite of some ongoing debate.Several other vital theories have been proposed from time to time but most of them were discarded.

All those theories which consider dead cells of xylem to be responsible for ascent of sap are grouped under physical theories.Several physical theories mainly based on physical forces such as imbibitions, capillary force,atmospheric pressure etc.

1.17. The Cohesion–Tension Theory (Transpiration pull and cohesion of water Theory) Explains Water Transport in the Xylem

Theory was proposed by Dixon and Joly. The theory is based on the following concepts-

1. Cohesive and adhesive properties of water molecule to form an unbroken continuous water column in the xylem.
2. Transpiration pull or tension exerted on this water column.

In theory, the pressure gradients needed to move water through the xylem could result from the generation of positive pressures at the base of the plant or

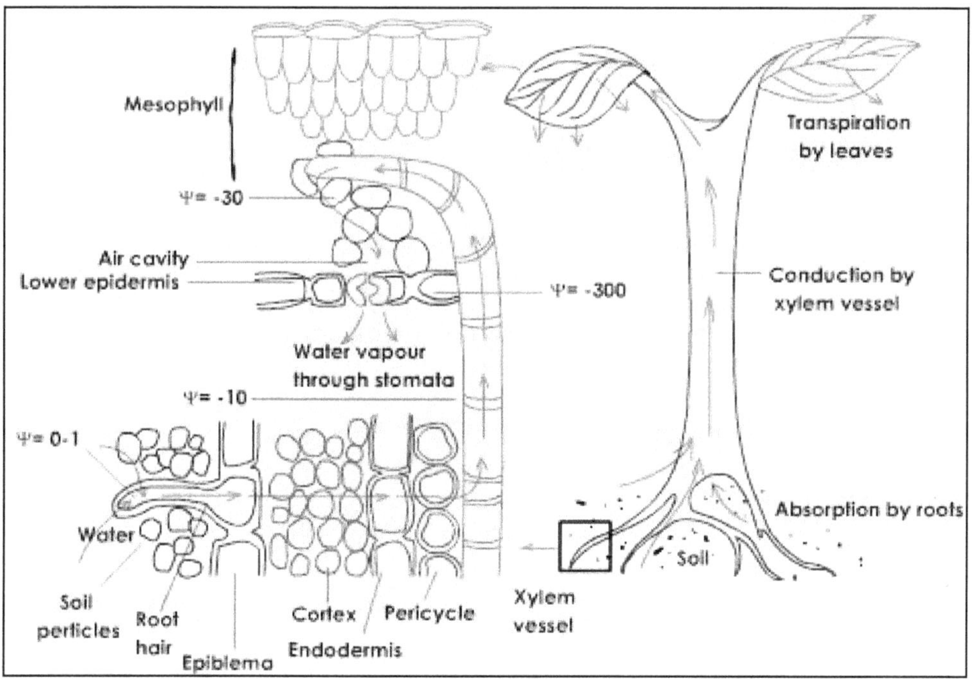

Figure 1.10: Diagrammatic Representation of Path of Water.

negative pressures at the top of the plant. Roots can develop positive hydrostatic pressure in their xylem–the so-called root pressure. However, root pressure is typically less than 0.1 MPa and disappears when the transpiration rate is high, so it is clearly inadequate to move water up a tall tree. Instead, the water at the top of a tree develops a large tension (a negative hydrostatic pressure), and this tension *pulls* water through the xylem. This mechanism, first proposed toward the end of the nineteenth century, is called the cohesion–tension theory of sap ascent because it requires the cohesive properties of water to sustain large tensions in the xylem water columns. The main controversy surrounds the question of whether water columns in the xylem can sustain the large tensions (negative pressures) necessary to pull water up tall trees. The most recent debate began when researchers modified the cell pressure probe technique to be able to measure directly the tension in xylem vessels (Balling and Zimmermann, 1990 t). One can readily demonstrate xylem tensions by puncturing intact xylem through a drop of ink on the surface of a stem from a transpiring plant. When the tension in the xylem is relieved, the ink is drawn instantly into the xylem, resulting in visible streaks along the stem.

1.18. Transpiration

Most of the water absorbed byplants is ultimately lost into the atmosphere. Only a very small fraction of water(generally less than 5 per cent) is utilized in plant development and metabolic process. The loss of water in the form of vapours from the living tissue of aerial parts of the plants is termed as transpiration.

Types of Respiration

1. Stomatal transpiration
2. Cuticular transpiration
3. Lenticular transpiration

1.18.1. Stomatal Transpiration

The stomata are tiny pores present in the epidermal surface of leaves, young stems and in certain fruits(*e.g.*- banana,citrus,cucumber etc,).They are also found in almost all the young aerial parts of plants. A typical stoma is consists of two kidney shaped guard cells surrounding a pore.The guard cell is much smaller than other epidermal cells.they are,therefore, rapidally affected even in small change in turgor. In some species the two guard cells are surrounded by accessory or subsidiary cells which differ in shape and size from other epidermal cells. Gourd cells generally have thick walls towards pore and thin walls on opposite side. Moreover, the cellulose micelles in guard cells wall are oriented radially rather than laterally. The guard cell walls have elastiv properties.

The guard cells differ from other epidermal cells in mainly two aspects- 1. They contain chloroplast and perform phenomenon of photosynthesis, but lack rubisco and $NADP^+$ linked triose phosphate dehydrogenase enzyme, thus carbon reduction cycleis absent and 2. They have special type of wall thickenings.

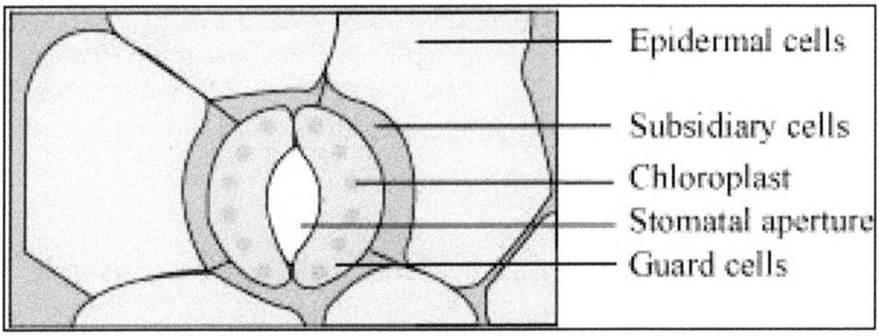

Figure 1.12: Structure of Stomata.

Functions

CO_2 Gain and Water Loss

The effectiveness of plants in moderating water loss while allowing sufficient CO_2 uptake for photosynthesis can be assessed by a parameter called the transpiration ratio. This value is defined as the amount of water transpired by the plant, divided by the amount of carbon dioxide assimilated by photosynthesis. For typical plants in which the first stable product of carbon fixation is a three-carbon compound, about 500 molecules of water are lost for every molecule of CO_2 fixed by photosynthesis, giving a transpiration ratio of 500. (Sometimes the reciprocal of the transpiration ratio, called the *water use efficiency*, is cited. Plants with a transpiration ratio of 500 have a water use efficiency of 1/500, or 0.002.) The large ratio of H_2O efflux to CO_2 influx results from three factors:

1. The concentration gradient driving water loss is about 50 times larger than that driving the influx of CO_2. In large part, this difference is due to the low concentration of CO_2 in air (about 0.03 per cent) and the relatively high concentration of water vapor within the leaf.
2. CO_2 diffuses about 1.6 times more slowly through air than water does (the CO2 molecule is larger than H_2O and has a smaller diffusion coefficient).
3. CO_2 uptake must cross the plasma membrane, the cytoplasm, and the chloroplast envelope before it is assimilated in the chloroplast. These membranes add to the resistance of the CO_2 diffusion pathway.

Alternative Approaches

Ordinarily, carbon dioxide is fixed to ribulose-1,5-bisphosphate (RuBP) by the enzyme rubisco in mesophyll cells exposed directly to the air spaces inside the leaf. This exacerbates the transpiration problem for two reasons: first, RuBisCo has a relatively low affinity for carbon dioxide, and second, it fixes oxygen to RuBP, wasting energy and carbon in a process called photorespiration. For both of these reasons, rubisco needs high carbon dioxide concentrations, which means wide stomatal apertures and, as a consequence, high water loss.

Narrower stomatal apertures can be used in conjunction with an intermediary molecule with a high carbon dioxide affinity, PEPcase (Phosphoenolpyruvate carboxylase). Retrieving the products of carbon fixation from PEPCase is in an energy-intensive process, however. As a result, the PEPCase alternative is preferable only where water is limiting but light is plentiful, or where high temperatures increase the solubility of oxygen relative to that of carbon dioxide, magnifying RuBisCo's oxygenation problem.

CAM Plants

A group of mostly desert plants called "CAM" plants (Crassulacean acid metabolism, after the family Crassulaceae, which includes the species in which the CAM process was first discovered) open their stomata at night (when water evaporates more slowly from leaves for a given degree of stomatal opening), use PEPcarboxylase to fix carbon dioxide and store the products in large vacuoles. The following day, they close their stomata and release the carbon dioxide fixed the previous night into the presence of rubisco. This saturates rubisco with carbon dioxide, allowing minimal photorespiration. This approach, however, is severely limited by the capacity to store fixed carbon in the vacuoles, so it is preferable only when water is severely limiting.

Opening and Closing of Stomata (Stomatal Movement)

A variety of external stimuli affects the opening and closing of stomata, the some among tem are light and dark, CO_2 concentration, water supply, pH of the cell sap etc.

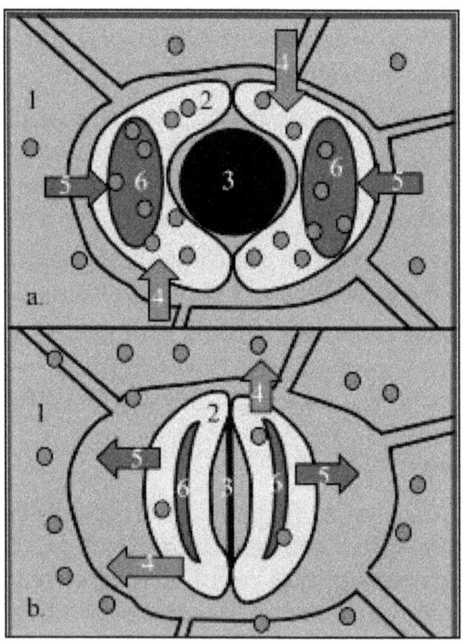

Figure 1.13: An Open Stoma (a) and a Closed Stoma (b)
(1: Epidermal cell; 2: Guard cell; 3: Stoma; 4: K^+ ions; 5: Water; 6: Vacuole)

However, most plants do not have the aforementioned facility and must therefore open and close their stomata during the daytime in response to changi conditions, such as light intensity, humidity, and carbon dioxide concentration. It is not entirely certain how these responses work. However, the basic mechanism involves regulation of osmotic pressure.When conditions are conducive to stomatal opening (*e.g.*, high light intensity and high humidity), a proton pump drives protons (H+) from the guard cells. This means that the cells' electrical potential becomes increasingly negative. The negative potential opens potassium voltage-gated channels and so an uptake of potassium ions (K^+) occurs. To maintain this internal negative voltage so that entry of potassium ions does not stop, negative ions balance the influx of potassium. In some cases, chloride ions enter, while in other plants the organic ion malate is produced in guard cells. This increase in solute concentration lowers the water potential inside the cell, which results in the diffusion of water into the cell through osmosis. This increases the cell's volume and turgor pressure. Then, because of rings of cellulose microfibrils that prevent the width of the guard cells from swelling, and thus only allow the extra turgor pressure to elongate the guard cells, whose ends are held firmly in place by surrounding epidermal cells, the two guard cells lengthen by bowing apart from one another, creating an open pore through which gas can move. When the roots begin to sense a water shortage in the soil, abscisic acid (ABA) is released. ABA binds to receptor proteins in the guard cells' plasma membrane and cytosol, which first raises the pH of the cytosol of the cells and cause the concentration of free Ca^{2+} to increase in the cytosol due to influx from outside the cell and release of Ca^{2+} from internal stores such as the endoplasmic reticulum and vacuoles. This causes the chloride (Cl^-) and inorganic ions to exit the cells. Second, this stops the uptake of any further K^+ into the cells and, subsequently, the loss of K+. The loss of these solutes causes an increase in water potential, which results in the diffusion of water back out of the cell by osmosis. This makes the cell plasmolysed, which results in the closing of the stomatal pores. Guard cells have more chloroplasts than the other epidermal cells from which guard cells are derived.

Factors Affecting the Stomatal Movement

1. Low CO_2 cocentration in the intercellular spaces of leaves causes stomata to open.On the other hand, a high concentration of CO_2 in leaves causes closing to stomata.
2. A turgid leaf in the dark,with closed stomata,shows stomatal opening shortly after exposure to light.
3. Stomata remaining closed at 0°C or lower than 0°C even under continous light.Rise in temperature upto 30°C causes stomatal opening.
4. Under condition of water deficit (increased water stress or decreased water potential) within a leaf, the stomata of many mesophytes get closed.
5. Rise in the pH of guard cells results opening of stomata and decrease in the pH causes its closure.
6. ABA plant hormone causes stomatal aperture to close.

Mechanism of Stomatal Movement

Active Potassium Theory (Levitt 1974)

It was observed by Fujino (1967) that opening of stomata occurs due to the influx of K^+ ions into the guard cells. The source of K^+ ions are the neighbouring subsidiary and epidermal cells there by increasing the concentration from 50mM to 300mM in guard cells.It has been proved that the accumulation of K^+ ions brings the opening of stomata and loss of K^+ ions, the closing stomata.

During day light

↓

Accumulation of K^+ ions by the guard cells

↓

Increased solute concentration

↓

Decreases water potential

↓

Endosmosis of water

↓

Increased turgidity

↓

Stoma open

During night/dark

↓

Loss of K + ions by guards cell

↓

Decreased solute concentration

↓

Increased water potential

↓

Exosmosis of water

↓

Dcreased turgidity

↓

Stoma closed

Figure 1.14: The Increase in K^+ Ion Concentration Increases the Osmotic Concentration of Guard Cells thus Leading to Stomatal Opening. ATP helps in entry of K^+ions into the guard cells. (Courtesy of Hopkins).

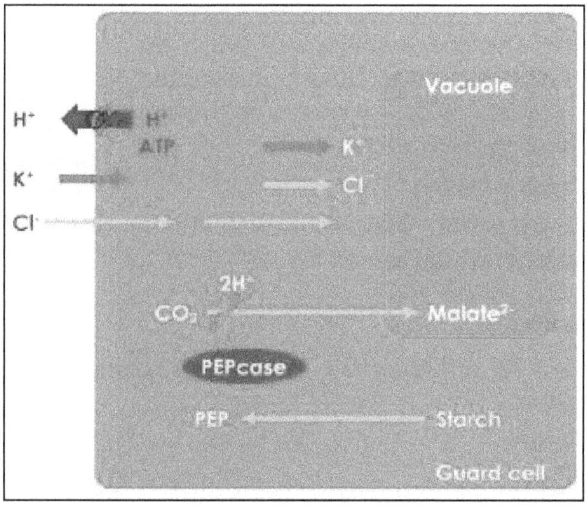

Figure 1.15: Role of Potassium, Chloride and Malate Ions in Stomatal Opening. (PEPcase = Phosphoenol pyruvate carboxylase).

Levitt (1974) observed that proton (H^+) uptake by guard cells, chloroplasts takes place with the help of ATP. This leads to increase in value of pH in guard cells. Rise in pH converts starch into organic acid like malic acid.

$$\text{Starch} \longrightarrow \text{Phosphoglyceric acid} \xrightarrow{\text{carboxylase}} \text{Phosphoenol pyruvic acid} + CO_2$$

$$\downarrow$$

$$\text{Malic acid} \longleftarrow \text{Oxaloacetic acid}$$

The uptake of K^+ ions is balanced by:

- ☆ Uptake of chloride (Cl^-) ions
- ☆ Transport of H^+ ions released from organic acid (malic acid)
- ☆ By negative charges of organic acids when they lose H^+ ions

Thus all these factors lead to the opening of stomata. The stomata closure is due to excretion of K^+ ions from guard cells surrounding epidermal and subsidiary cells.The stomatal closure is considered to be brought about by a passive or highly catalysed excretion of K^+ ions and Cl^- ions from the guard cells to the epidermal tissue in general and the subsidiary cells in particular. It is believed that subsidiary cells have an active reabsorption mechanism of K^+ ions.

Factors Affecting the Rate of Transpiration

(A) External Factors

1. Atmospheric humidity - The rate of transpiration decreases with increases in relative humidity of atmosphere.
2. Temperature – The rate of transpiration increases with the increases in atmospheric temperature.
3. Light – It indirectly affects the rate of transpiration by opening the stomata and increasing the temperature.

4. Wind velocity – A transpiring surface of leaf continuously adds water vapours to the atmospheric air. Once the immediate area becomes saturated, it reduces the rate of transpiration. Wind velocity removes the air of that area, which is replaced by fresh air and results in an increase in the rate of transpiration.
5. Atmospheric pressure
6. Available soil water – If available water in the soil is not sufficient the rate of transpiration is decreased.

(B) Internal Factors

1. Leaf area – By reducing the size of leaf, plants decrease their rate of transpiration.
2. Leaf structure
3. Age of plants

To avoid the water loss find out the substances that can reduce the transpiration without affecting the other processes, therefore, a substance applied to the plant for the purpose of reducing transpiration, known as antitranspirant. Several substances such as – colour less plastics, silicone oils, low viscosity waxes, fungicide phenylmercuric acetate, ABA etc.

The advantages of transpiration are – 1. It increase the rate of absorption of water, 2- It helps in the mineral absorption and ascent of sap, 3- It regulate the plant temperature.

Guttation

The process of exudation of liquid drops from the edge of leaves is called guttation.Usually it occurs through stomata like pores called hydathodes. The cause of guttation is mainly root pressure. Pores are present over amass of loosely arranged cells with large intercellular spaces called epithem. It ceases during the day and nearly always occure at night. Mineral deficiency in plants may inhibit guttation. It is uncontrolled phenomenon.

References

Borstlap, A.C. (2002). Early diversification of plant aquaporins. *Trends in Plant Science*, 7: 529-530.

Maurel, C.L., Verdoucq, D.T. Luu, and Santoni, V. (2008). Plant aquaporins: membrane channels with multiple integrated functions. *Annual Review of Plant Biology*, 59: 596-624.

Tanner, W. and Beevers, H. (2001). Transpiration, a prerequisite for long distance transport of minerals in plants. *Proceeding of the National Academy of Sciences U.S.A.* 98: 9443-9447.

Tibbettts, T.J. and Ewers, F.W. (2000). Root pressure and specific conductivity in temperate lianas: Exotic *Celastrus orbiculatus* (Celastraceae) vs. native *Vitis riparia* (Vitaceae). *American Journal of Botany*, 87: 1272-1278.

Chapter 2

Mineral Nutrition

Plants require some inorganic matter from the soil, where they are present in the form of minerals, they are called mineral elements or mineral nutrients.

Only certain elements have been determined to be essential for plant growth. There are two principal criteria can be judged essential or nonessential to any plant –

1. An element is essential if the plant cannot complete its life cycle, (*i.e.* from viable seeds) in the absence of that element.
2. An element is essential if it forms part of any molecule or constituent of the plant that is itself essential in the plant (*e.g.*, N in protein and chlorophyll).

An essential element is defined as one whose absence prevents a plant from completing its life cycle (Arnon and Stout 1939) or one that has a clear physiological role (Epstein, 1999). If plants are given these essential elements, as well as energy from sunlight, they can synthesize all the compounds they need for normal growth. Table 1.0 lists the elements that are considered to be essential for most, if not all, higher plants. The first three elements–hydrogen, carbon, and oxygen–are not considered mineral nutrients because they are obtained primarily from water or carbon dioxide

The micronutrient cations iron, and to a lesser extent zinc, manganese and copper, are relatively insoluble in nutrient solution and most soil solution. Under these conditions, the micronutrient cations react with hydroxyl ions eventually precipitating insoluble hydrous metal oxides. Certain plants, especially cereals grains and other grasses cannot absorb enough iron and zinc. One way to overcome this defiency is to fertilize with metal chelates. A chelate is the soluble product formed when certain atoms in an organic molecule called a chelting agent or ligands donate electrons to the metal cation. Negatively charged carboxyl groups and nitrogen atoms possess electrons that can be shared in this way. One best known synthetic ligands is EDTA (ethylene-diaminetetraacetic acid).

Table 2.1

Element	Chemical Symbol	Concentration in Dry Matter (Per cent or ppm)	Relative Number of Atoms with Respect to Molybdenum
Obtained from water or carbon dioxide			
Hydrogen	H	6	60,000,000
Carbon	C	45	40,000,000
Oxygen	Q	45	30,000,000
Obtained from the soil			
Macronutrients			
Nitrogen	N	1.5	1,000,000
Potassium	K	1.0	250,000
Calcium	Ca	0.5	125,000
Magnesium	Mg	0.2	80,000
Phosphorus	P	0.2	60,000
Sulfur	S	0.1	30,000
Silicon	Si	0.1	30,000
Micronutrients			
Chlorine	Cl	100	3,000
Iron	Fe	100	2,000
Boron	B	20	2,000
Manganese	Mn	50	1,000
Sodium	Na	10	400
Zinc	Zn	20	300
Copper	Cu	6	100
Nickel	Ni	0.1	2
Molybdenum	Mo	0.1	1

Naturally occurring elements, other than those listed in Table 2.1, can also accumulate in plant tissues. For example, aluminum is not considered to be an essential element, but plants commonly contain from 0.1 to 500 ppm aluminum, and addition of low levels of aluminum to a nutrient solution may stimulate plant growth (Marschner 1995). Many species in the genera *Astragalus, Xylorhiza, Stanleya, Maize and Equisetum arvense* accumulate selenium, although plants have not been shown to have a specific requirement for this element. Cobalt is part of cobalamin (vitamin B12 and its derivatives), a component of several enzymes in nitrogen-fixing microorganisms. Thus cobalt deficiency blocks the development and function of nitrogen-fixing nodules. Nonetheless, plants that do not fix nitrogen, as well as nitrogen-fixing plants that are supplied with ammonium or nitrate, do not require cobalt.Some plants require nickel, because Ni^{2+} is an essential part of the enzyme urease. Crop plants normally contain only relatively small amounts of nonessential elements.

In the nineteenth century, several researchers, including Nicolas-Théodore de Saussure, Julius von Sachs, Jean-Baptiste-Joseph-Dieudonné Boussingault, and Wilhelm Knop, approached this problem by growing plants with their roots immersed in a nutrient solution containing only inorganic salts. Their demonstration that plants could grow normally with no soil or organic matter proved unequivocally that plants can fulfill all their needs from only inorganic elements and sunlight. The technique of growing plants with their roots immersed in nutrient solution without soil is called solution culture or hydroponics (Gericke, 1937). Successful hydroponic culture requires a large volume of nutrient solution or frequent adjustment of the nutrient solution to prevent nutrient uptake by roots from producing radical changes in nutrient concentrations and pH of the medium. A sufficient supply of oxygen to the root system–also critical–may be achieved by vigorous bubbling of air through the medium. Hydroponics is used in the commercial production of many greenhouse crops.

2.1. Nutrient Deficiency Symptoms and Specific Function of Macro and Micronutrients

Group 1: Deficiencies in mineral nutrients that are part of carbon compounds. This first group consists of nitrogen and sulfur. Nitrogen availability in soils limits plant productivity in most natural and agricultural ecosystems. By contrast, soils generally contain sulfur in excess. Nonetheless, nitrogen and sulfur share the property that their oxidation–reduction states range widely. Some of the most energy-intensive reactions in life convert the highly oxidized, inorganic forms absorbed from the soil into the highly reduced forms found in organic compounds such as amino acids.

Nitrogen

Nitrogen is the mineral element that plants require in greatest amounts. It serves as a constituent of many plant cell components, including amino acids and nucleic acids. Nitrogen are absorbed in forms of – nitrate (NO_3^-) and ammonium (NH_4^+). Therefore, nitrogen deficiency rapidly inhibits plant growth. If such a deficiency persists, most species show chlorosis (yellowing of the leaves), especially in the older leaves near the base of the plant (for pictures of nitrogen deficiency and the other mineral deficiencies described. Under severe nitrogen deficiency, these leaves become completely yellow (or tan) and fall off the plant. Younger leaves may not show these symptoms initially because nitrogen can be mobilized from older leaves. Thus a nitrogen-deficient plant may have light green upper leaves and yellow or tan lower leaves. When nitrogen deficiency develops slowly, plants may have markedly slender and often woody stems. This woodiness may be due to a buildup of excess carbohydrates that cannot be used in the synthesis of amino acids or other nitrogen compounds. Carbohydrates not used in nitrogen metabolism may also be used in anthocyanin synthesis, leading to accumulation of that pigment. This condition is revealed as a purple coloration in leaves, petioles, and stems of some nitrogen-deficient plants, such as tomato and certain varieties of corn.Excess nitrogen also causes negative impact on some plants such as flowering and formation of seeds become retarded.

Sulfur

Sulfur is absorbed from soils as divalent sulphate anions (SO_4^{2-}). Sulfur is found in two amino acids (cysteine and methionine) and is a constituent of several coenzymes and vitamins (Thiamine and biotin and co-enzyme A) essential for Metabolism *i.e.* for respiration and for synthesis and breakdown of fatty acids. Many of the symptoms of sulfur deficiency are similar to those of nitrogen deficiency, including chlorosis, stunting of growth, and anthocyanin accumulation. This similarity is not surprising, since sulfur and nitrogen are both constituents of proteins. However, the chlorosis caused by sulfur deficiency generally arises initially in mature and young leaves, rather than in the old leaves as in nitrogen deficiency, because unlike nitrogen, sulfur is not easily remobilized to the younger leaves in most species. Nonetheless, in many plant species sulfur chlorosis may occur simultaneously in all leaves or even initially in the older leaves.

Group 2: Deficiencies in mineral nutrients that are important in energy storage or structural integrity. This group consists of phosphorus, silicon, and boron. Phosphorus and silicon are found at concentrations within plant tissue that warrant their classification as macronutrients, whereas boron is much less abundant and considered a micronutrient. These elements are usually present in plants as ester linkages to a carbon molecule.

Phosphorus

It is absorbed primarily as the monovalent phosphate anion (PO_4^-) and less rapidly as the divalent anion (PO_4^{2-}). Phosphate (as phosphate, PO_4^{3-}) is an integral component of important compounds of plant cells, including the sugar–phosphate intermediates of respiration and photosynthesis, and the phospholipids that make up plant membranes. It is also a component of nucleotides used in plant energy metabolism (such as ATP) and in DNA and RNA. Characteristic symptoms of phosphorus deficiency include stunted growth in young plants and a dark green coloration of the leaves, which may be malformed and contain small spots of dead tissue called necrotic spots. As in nitrogen deficiency, some species may produce excess anthocyanins, giving the leaves a slight purple coloration. In contrast to nitrogen deficiency, the purple coloration of phosphorus deficiency is not associated with chlorosis. In fact, the leaves may be a dark greenish purple. Additional symptoms of phosphorus deficiency include the production of slender (but not woody) stems and the death of older leaves. Maturation of the plant may also be delayed. If excess phosphorus is provided, root growth is often increased relative to shoot growth.This in contrast to effects with excess nitrogen, causes low shoot- to- root ratio.

Silicon

Silicon exists in soil solutions as silic acid,(H_4SiO_4) and absorbed in this form. Only members of the family Equisetaceae–called *scouring rushes* because at one time their ash, rich in gritty silica, was used to scour pots–require silicon to complete their life cycle. Nonetheless, many other species accumulate substantial amounts of silicon within their tissues and show enhanced growth and fertility when supplied with adequate amounts of silicon (Epstein 1999). Plants deficient in silicon are more susceptible to lodging (falling over) and fungal infection. Silicon is deposited primarily in the endoplasmic reticulum, cell walls, and intercellular spaces as hydrated, amorphous silica ($SiO_2 \cdot nH_2O$). It also forms complexes with polyphenols and thus serves as an alternative to lignin in the reinforcement of cell walls. In addition, silicon can ameliorate the toxicity of many heavy metals.

Boron

Boron is almost entirely absorbed from soils as undissociate boric acid (H_3BO_3). Although the precise function of boron in plant metabolism is unclear, evidence suggests that it plays roles in cell elongation, nucleic acid synthesis, hormone responses, and membrane function (Shelp 1993). Boron deficient plants may exhibit a wide variety of symptoms, depending on the species and the age of the plant. A characteristic symptom is black necrosis of the young leaves and terminal buds. The necrosis of the young leaves occurs primarily at the base of the leaf blade. Stems may be unusually stiff and brittle. Apical dominance may also be lost, causing the plant to become highly branched; however, the terminal apices of the branches soon become necrotic because of inhibition of cell division. Structures such as the fruit, fleshy roots, and tubers may exhibit necrosis or abnormalities related to the breakdown of internal tissues.some common deficiency disorder are- 'Heart rot' of beets, 'stem crack' of celery, 'water core' of turnip and 'drought spot' of apples

***Group** 3*: Deficiencies in mineral nutrients that remain in ionic form. This group includes some of the most familiar mineral elements: The macronutrients potassium, calcium, and magnesium, and the micronutrients chlorine, manganese, and sodium. They may be found in solution in the cytosol or vacuoles, or they may be bound electrostatically or as ligands to larger carbon-containing compounds.

Potassium

After nitrogen and phosphorus, soils are usually most deficient in potassium. Potassium is rich in actively dividing cells of buds,young leaves and root tips. Potassium, present within plants as the cation K^+, plays an important role in regulation of the osmotic potential of plant cells. It also activates many enzymes involved in respiration and photosynthesis.It regulates movement of stomata. The first observable symptom of potassium deficiency is mottled or marginal chlorosis, which then develops into necrosis primarily at the leaf tips, at the margins, and between veins. In many monocots, these necrotic lesions may initially form at the leaf tips and margins and then extend toward the leaf base. Because potassium can be mobilized to the younger leaves, these symptoms appear initially on the more mature leaves toward the base of the plant. The leaves may also curl and crinkle. The stems of potassium-deficient plants may be slender and weak, with abnormally short internodal regions. In potassium-deficient corn, the roots may have an increased susceptibility to root-rotting fungi present in the soil, and this susceptibility, together with effects on the stem, results in an increased tendency for the plant to be easily bent to the ground (lodging).

Calcium

Calcium is an exchangeable cation. Calcium ions (Ca^{2+}) are used in the synthesis of new cell walls, particularly the middle lamellae that separate newly divided cells. Calcium is also used in the mitotic spindle during cell division. It is required for the normal functioning of plant membranes and has been implicated as a second messenger for various plant responses to both environmental and hormonal signals (Sanders *et al.*, 1999). In its function as a second messenger, calcium may bind to calmodulin, a protein found in the cytosol of plant cells. The calmodulin–calcium complex regulates many metabolic processes.In many plants it is also found as insoluble crystal of calcium oxalate.Calcium is almost immobile in phloem and as a defiency symptoms are always more pronounced in young tissue. Characteristic symptoms of calcium deficiency include necrosis of young meristematic regions, such as the tips of roots or young leaves, where cell division and wall formation are most rapid. Necrosis in slowly growing plants may be preceded by a general chlorosis and downward hooking of the young leaves. Young leaves may also appear deformed. The root system of a calcium-deficient plant may appear brownish, short, and highly branched. Severe stunting may result if the meristematic regions of the plant die prematurely.

Magnesium

In plant cells, magnesium ions (Mg^{2+}) have a specific role in the activation of enzymes involved in respiration, photosynthesis, and the synthesis of DNA and RNA. Magnesium is also a part of the ring structure of the chlorophyll molecule. A characteristic symptom of magnesium deficiency is chlorosis between the leaf veins, occurring first in the older leaves because of the mobility of this element. This pattern of chlorosis results because the chlorophyll in the vascular bundles remains unaffected for longer periods than the chlorophyll in the cells between the bundles does. If the deficiency is extensive, the leaves may become yellow or white. An additional symptom of magnesium deficiency may be premature leaf abscission.

Chlorine

The element chlorine is found in plants as the chloride ion (Cl^-). It is required for the water-splitting reaction of photosynthesis through which oxygen is produced (Clarke and Eaton-Rye, 2000). In addition,

chlorine may be required for cell division in both leaves and roots (Harling *et al.*, 1997). Plants deficient in chlorine develop wilting of the leaf tips followed by general leaf chlorosis and necrosis. The leaves may also exhibit reduced growth. Eventually, the leaves may take on a bronzelike color ("bronzing"). Roots of chlorine-deficient plants may appear stunted and thickened near the root tips. Chloride ions are very soluble and generally available in soils because seawater is swept into the air by wind and is delivered to soil when it rains. Therefore, chlorine deficiency is unknown in plants grown in native or agricultural habitats. Most plants generally absorb chlorine at levels much higher than those required for normal functioning.

Manganese

Manganese ions (Mn^{2+}) activate several enzymes in plant cells. In particular, decarboxylases and dehydrogenases involved in the tricarboxylic acid (Krebs) cycle are specifically activated by manganese. The best defined function of manganese is in the photosynthetic reaction through which oxygen is produced from water (Marschner, 1995), nitrogen metabolism and oxidation of auxin. The major symptom of manganese deficiency is intervenous chlorosis associated with the development of small necrotic spots. This chlorosis may occur on younger or older leaves, depending on plant species and growth rate.The various disorders such as - "grey speck" of oats, "marsh spot" of peas and "speckled yellows" of sugar beets results when inadequate amounts are presents. In spinch leaves the absence of Mn causes disorganisation of thylakoid membranes.

Sodium

Most species utilizing the C4 and CAM pathways of carbon fixation require sodium ions (Na^{+}). In these plants, sodium appears vital for regenerating phosphoenolpyruvate, the substrate for the first car-boxylation in the C4 and CAM pathways (Johnstone *et al.*, 1988).It regulates the transport of amino acids to the nucleus. Under sodium deficiency, these plants exhibit chlorosis and necrosis, or even fail to form flowers. Many C3 species also benefit from exposure to low levels of sodium ions. Sodium stimulates growth through enhanced cell expansion, and it can partly substitute for potassium as an osmotically active solute.

Group 4: Deficiencies in mineral nutrients that are involved in redox reactions.

This group of five micronutrients includes the metals iron, zinc, copper, nickel, and molybdenum. All of these can undergo reversible oxidations and reductions (*e.g.*, Fe^{2+}~Fe^{3+}) and have important roles in electron transfer and energy transformation. They are usually found in association with larger molecules such as cytochromes, chlorophyll, and proteins (usually enzymes).

Iron

Iron is absorbed in ferrous form. The soil pH controls the iron availability to plants. Iron has an important role as a component of enzymes involved in the transfer of electrons (redox reactions), such as cytochromes. In this role, it is reversibly oxidized from Fe^{2+} to Fe^{3+} during electron transfer. As in magnesium deficiency, a characteristic symptom of iron deficiency is intervenous chlorosis. In contrast to magnesium deficiency symptoms, these symptoms appear initially on the younger leaves because iron cannot be readily mobilized from older leaves. Under conditions of extreme or prolonged deficiency, the veins may also become chlorotic, causing the whole leaf to turn white. The leaves become chlorotic because iron is required for the synthesis of some of the chlorophyll–protein complexes in the chloroplast. The low mobility of iron is probably due to its precipitation in the older leaves as insoluble oxides or phosphates or to the formation of complexes with phytoferritin, an iron-binding protein found in the leaf and other plant parts (Oh *et al.*, 1996). Phytoferritin seems to represents a storehouse of iron. The precipitation of iron diminishes subsequent mobilization of the metal into the phloem for long-distance translocation.

Zinc

Many enzymes require zinc ions (Zn^{2+}) for their activity, and zinc may be required for chlorophyll biosynthesis in some plants. Zn play important role in synthesis of tryptophan which is precursor of auxin hormone. Zinc deficiency is characterized by a reduction in internodal growth, and as a result plants display a rosette habit of growth in which the leaves form a circular cluster radiating at or close to the ground. The leaves may also be small and distorted e,g,- "little leaf"disease of brinjal, "rossete" of apples, peaches with leaf margins having a puckered appearance. These symptoms may result from loss of the capacity to produce sufficient amounts of the auxin indole acetic acid. In some species (corn, sorghum, beans), the older leaves may become intervenously chlorotic and then develop white necrotic spots. This chlorosis may be an expression of a zinc requirement for chlorophyll biosynthesis.

Copper

A very little amounts is found dissolved in the soil solution.It is toxic when presnt in larger amount. Like iron, copper is associated with enzyme involved in redox reactions being reversibly oxidized from Cu^{+} to Cu^{2+}. An example of such an enzyme is plastocyanin, which is involved in electron transfer during the light reactions of photosynthesis (Haehnel, 1984).It is a component of enzye involved in the synthesis of ascorbic acid (vit. C). The initial symptom of copper deficiency is the production of dark green leaves, which may contain necrotic spots. The necrotic spots appear first at the tips of the young leaves and then extend toward the leaf base along the margins. The leaves may also be twisted or malformed. Under extreme copper deficiency, leaves may abscise prematurely. Citrus orchards are occasionally deficient in which case the dying young leaves led to name "die back" disease.

Nickel

Urease is the only known nickel-containing enzyme in higher plants, although nitrogen-fixing microorganisms require nickel for the enzyme that reprocesses some of the hydrogen gas generated during fixation (hydrogen uptake hydrogenase). Nickel deficient plants accumulate urea in their leaves and, consequently, show leaf tip necrosis. Plants grown in soil seldom, if ever, show signs of nickel deficiency because the amounts of nickel required are minuscule.

Molybdenum

Molybdenum ions (Mo^{4+} through Mo^{6+}) are components of several enzymes, including nitrate reductase and nitrogenase. Nitrate reductase catalyzes the reduction of nitrate to nitrite during its assimilation by the plant cell; nitrogenase converts nitrogen gas to ammonia in nitrogen-fixing microorganisms. The first indication of a molybdenum deficiency is general chlorosis between veins and necrosis of the older leaves. In some plants, such as cauliflower or broccoli, the leaves may not become necrotic but instead may appear twisted and subsequently d (whiptail disease). Flower formation may be prevented, or the flowers may abscise prematurely. Because molybdenum is involved with both nitrate assimilation and nitrogen fixation, a molybdenum deficiency may bring about a nitrogen deficiency if the nitrogen source is primarily nitrate or if the plant depends on symbiotic nitrogen fixation. Although plants require only small amounts of molybdenum, some soils supply inadequate levels. Small additions of molybdenum to such soils can greatly enhance crop or forage growth at negligible cost.

2.2. Absorption and Transport of Mineral Salts (or solute)

The soil is a complex physical, chemical, and biological substrate. It is a heterogeneous material containing solid, liquid, and gaseous phases. All of these phases interact with mineral elements. The inorganic particles of the solid phase provide a reservoir of potassium, calcium, magnesium, and iron. Hydrogen ion concentration (pH) is an important property of soils because it affects the growth of plant roots and soil microorganisms. Root growth is generally favored in slightly acidic soils, at pH values between 5.5 and 6.5. Fungi generally predominate in acidic soils; bacteria become more prevalent in alkaline soils. Soil pH determines the availability of soil nutrients. When excess minerals are present in the soil, the soil is said to be saline, and plant growth may be restricted if these mineral ions reach levels that limit water availability or exceed the adequate zone for a particular nutrient. The ability of plants to obtain both water and mineral nutrients from the soil is related to their capacity to develop an extensive root system. Plant roots may grow continuously throughout the year. Their proliferation, however, depends on the availability of water and minerals in the immediate microenvironment surrounding the root, the so-called rhizosphere. If the rhizosphere is poor in nutrients or too dry, root growth is slow. As rhizosphere conditions improve, root growth increases. Root hairs, with their large surface area for absorption of water and solutes, first appear in the maturation zone and it is here that the xylem develops the capacity to translocate substantial quantities of water and solutes to the shoot. When diffusion is too slow to maintain high nutrient concentrations near the root, a nutrient depletion zone forms adjacent to the root surface.

The direct acquisition of mineral elements by the root, but this process may be modified by the association of mycorrhizal fungi with the root system. Mycorrhizae (singular mycorrhiza, from the Greek words for "fungus" and "root") are not unusual; in fact, they are widespread under natural conditions. The root associated with mycorrhizal fungi: 83 per cent of dicots, 79 per cent of monocots, and all gymnosperms regularly form mycorrhizal associations (Wilcox 1991). On the other hand, plants from the families Cruciferae (cabbage), Chenopodiaceae (spinach), and Proteaceae (macadamia nuts), as well as aquatic plants, rarely if ever have mycorrhizae. Mycorrhizae are absent from roots in very dry, saline, or flooded soils, or where soil fertility is extreme, either high or low. Mycorrhizal fungi are composed of fine, tubular filaments called hyphae (singular hypha). The mass of hyphae that forms the body of the fungus is called the mycelium (plural mycelia). There are two major classes of mycorrhizal fungi: ectotrophic mycorrhizae and vesicular-arbuscular mycorrhizae (Smith *et al.*, 1997). Minor classes of mycorrhizal fungi include the ericaceous and orchidaceous mycorrhizae. Ectotrophic mycorrhizal

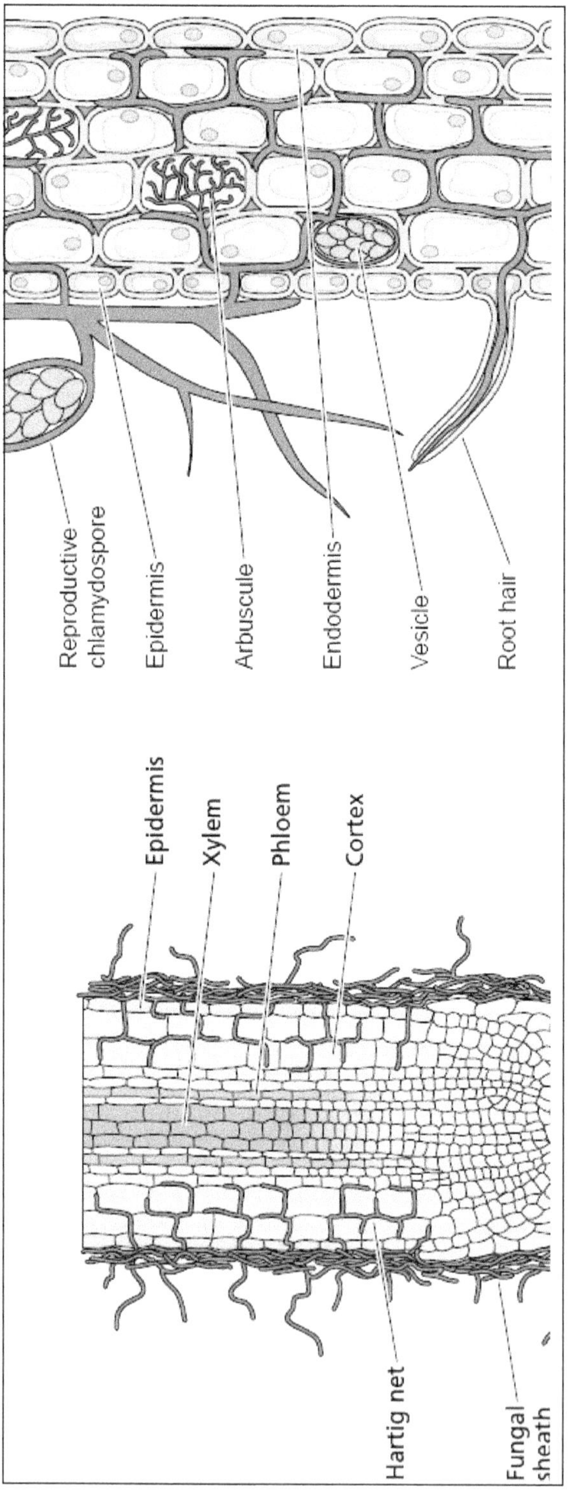

Figure 2.1: Root Affected Mycorrhizal Fungi, (A) Ectotrophic, (B) VAM.

fungi typically show a thick sheath, or "mantle," of fungal mycelium around the roots, and some of the mycelium penetrates between the cortical cells. The cortical cells themselves are not penetrated by the fungal hyphae but instead are surrounded by a network of hyphae called the Hartig net.

The capacity of the root system to absorb nutrients is improved by the presence of external fungal hyphae that are much finer than plant roots and can reach beyond the areas of nutrient-depleted soil near the roots (Clarkson 1985). Ectotrophic mycorrhizal fungi infect exclusively tree species, including gymnosperms and woody angiosperms. Unlike the ectotrophic mycorrhizal fungi, vesicular arbuscular mycorrhizal fungi do not produce a compact mantle of fungal mycelium around the root. Instead, the hyphae grow in a less dense arrangement, both within the root itself and extending outward from the root into the surrounding soil (Figure2.1). After entering the root through either the epidermis or a root hair, the hyphae not only extend through the regions between cells but also penetrate individual cells of the cortex. Within the cells, the hyphae can form oval structures called vesicles and branched structures called arbuscules. The arbuscules appear to be sites of nutrient transfer between the fungus and the host plant. The association of vesicular-arbuscular mycorrhizae with plant roots facilitates the uptake of phosphorus and trace metals such as zinc and copper. In return, plants provide carbohydrates to the mycorrhizae.

2.3. Mechanism of Mineral Absorption

Various theories have been put forward to explain the mechanism of solute absorption. They are boadly categories in to two groups, such as – 1. Passive, 2. Active absorption.

2.3.1. Passive Absorption (or Transport)

An absorption of ions and molecules without involving metabolic energy. According to Fick's first law, the movement of molecules by diffusion always proceeds spontaneously, down a gradient of concentration or chemical potential, until equilibrium is reached. The spontaneous "downhill" movement of molecules is termed passive transport. The various modes of passive absorption are:

2.3.2. Diffusion

The molecules or ions will diffuse into cells as long as is maintained the chemical potential gradient and until an equilibrium is estabilished. Thus the absorption of solute by process of diffusion is directly proportional to the driving force.

2.3.3. Facilitated Diffusion

The solute mediated by a carrier such as some protein, act as shuttles for a passive transport of ions or molecules across membrane. Certain polypeptides (small protein) produce by bacteria have been identified as carriers. These are called ionophores because they transport cations across the membranes.

2.3.4. Ion Exchange

The ions are charged, they can not be absorbed alone without upsetting the electrostatic equilibrium between positively and negatively charge particles.When positively charged ions such as K^+ is absorbed by a cell either positively charged ion such as H^+ is displaced from the cell (ion exchange) or a negatively charged on must accompany it.

2.3.5. Donnan Equilibrium

This describe the effects of fixed or non-diffusible anions which mostly accumulate on the inner side of outer membrane. Normally both anion and cation will diffuse into cells in equal numbers until an equilibrium between the cell sap and external medium is attained.

Mass flow of ions –

An increase in the rate of transpiration causes increase in the salt uptake.

2.3.6. Active Absorption

The movement of substances against or up a gradient of chemical potential (*e.g.*, to a higher concentration) is termed active transport. It is not spontaneous. An active transport is often called a pump, *e.g.* K^+ pump, Na^+ pump etc. An active transport is called uniport when one ion or molecule crosses a protoplasmic membrane at one site. When two substances cross it at the same site in the same direction it is called symport and when two cross a membrane at the same site in opposite direction it is called antiport.

2.4. Mechanism of Active Absorption or Transport

There are various possible mechanisms to explain solute transport across protoplasmic membrane or tonoplast.

2.4.1. Carrier Concept

carrier protein extend completely across the membrane. In transport mediated by a carrier, the substance being transported is initially bound to a specific site on the carrier protein. This requirement for binding allows carriers to be highly selective for a particular substrate to be transported. Carriers therefore specialize in the transport of specific organic metabolites. Binding causes a conformational change in the protein, which exposes the substance to the solution on the other side of the membrane. Transport is complete when the substance dissociates from the carrier's binding site.

2.4.2. Primary Active Transport

The membrane proteins that carry out primary active transport are called pumps. Most pumps transport ions, such as H^+ or Ca^{2+}. However, pumps belonging to the "ATP binding cassette" family of transporters can carry large organic molecules. Ion pumps can be further characterized as either electrogenic electroneutral. In general, electrogenic transport refers to ion transport involving the net movement of charge across the membrane. In the plasma membranes of plants, fungi, and

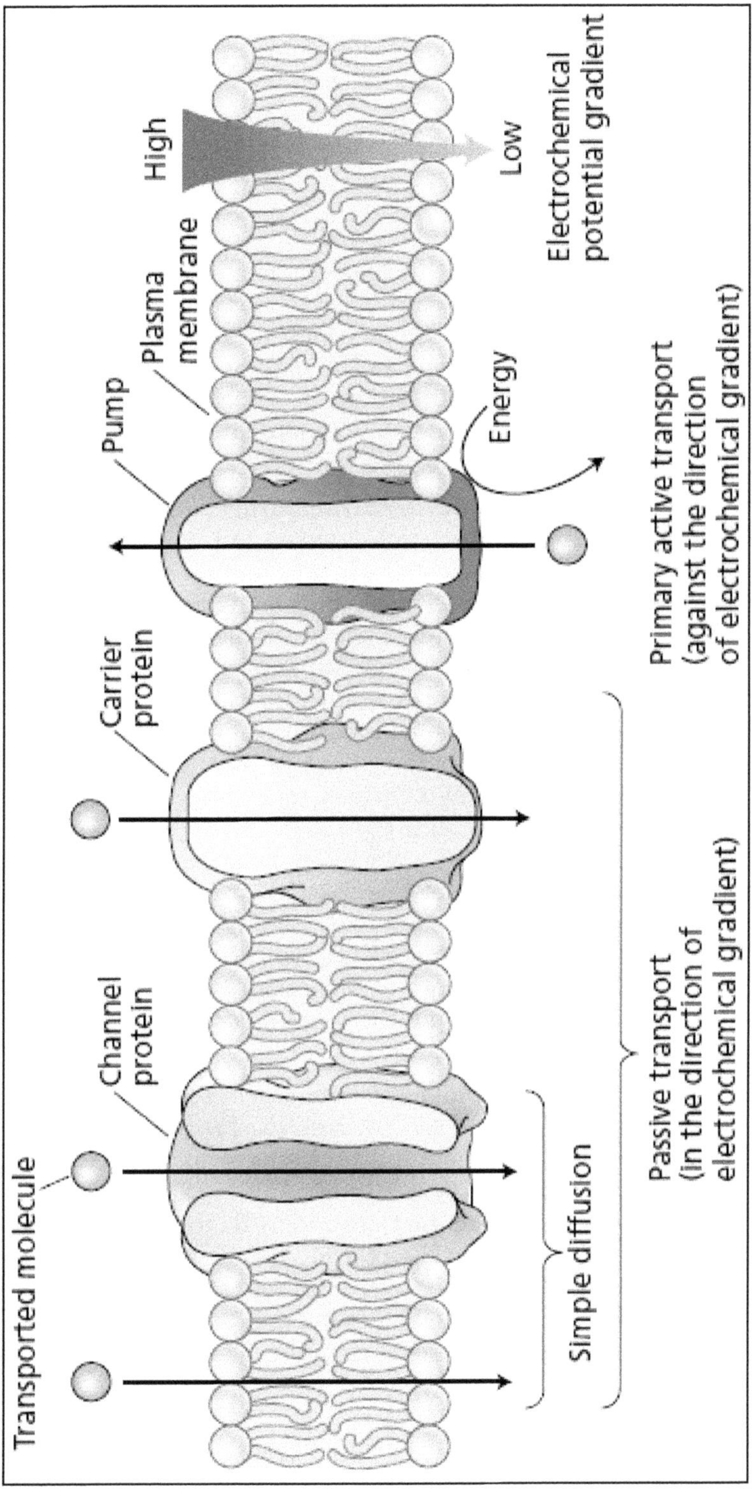

Figure 2.2: Classes of Membrane Transport.

bacteria, as well as in plant tonoplasts and other plant and animal endomembranes, H^+ is the principal ion that is electrogenically pumped across the membrane. The plasma membrane H^+-ATPase generates the gradient of electrochemical potentials of H^+ across the plasma membranes, while the vacuolar H^+-ATPase and the H^+-pyrophosphatase (H^+-PPase) electrogenically pump protons into the lumen of the vacuole and the Golgi cisternae.

2.4.3. Secondary Active Transport

In plant plasma membranes, the most prominent pumps are for H^+ and Ca^{2+}, and the direction of pumping is outward. Therefore another mechanism is needed to drive the active uptake of most mineral nutrients. The other important way that solutes can be actively transported across a membrane against their gradient of electrochemical potential is by coupling of the uphill transport of one solute to the downhill transport of another. This type of carriermediated cotransport is termed secondary active transport, and it is driven indirectly by pumps.

Protons are extruded from the cytosol by electrogenic H^+- ATPases operating in the plasma membrane and at the vacuole membrane. Consequently, a membrane potential and a pH gradient are created at the expense of ATP hydrolysis. This gradient of electrochemical potential for H, the proton motive force (PMF), or Δp, represents stored free energy in the form of the H^+ gradient. The proton motive force generated by electrogenic H^+ transport is used in secondary active transport to drive the transport of many other substances against their gradient of electrochemical potentials. Figure 2.2 shows how secondary transport may involve the binding of a substrate (S) and an ion (usually H^+) to a carrier protein, and a conformational change in that protein. There are two types of secondary transport: symport and antiport. The example shown in Figure 6.9 is called symport (and the protein involved is called a symporter) because the two substances are moving in the same direction through the membrane (see also Figure 2.3 A). Antiport (facilitated by a protein called an antiporter) refers to coupled transport in which the downhill movement of protons drives the active (uphill) transport of a solute in the opposite direction (Figure 2.3 B). In both types of secondary transport, the ion or solute being transported simultaneously with the protons is moving against its gradient of electrochemical potential, so its transport i

Typically, transport across a biological membrane is energized by one primary active transport system coupled to ATP hydrolysis. The transport of that ion–for example, H+–generates an ion gradient and an electrochemical potential. Many other ions or organic substrates can then be transported by a variety of secondary active-transport proteins, which energize the transport of their respective substrates by simultaneously carrying one or two H+ ions down their energy gradient. Thus H+ ions circulate across the membrane, outward through the primary active transport proteins, and back into the cell through the secondary transport proteins. In plants and fungi, sugars and amino acids are taken up by symport with protons.

Factors affecting Solute Absorption

Absorption of mineral salts is affected by the number of external and internal factors. They are:

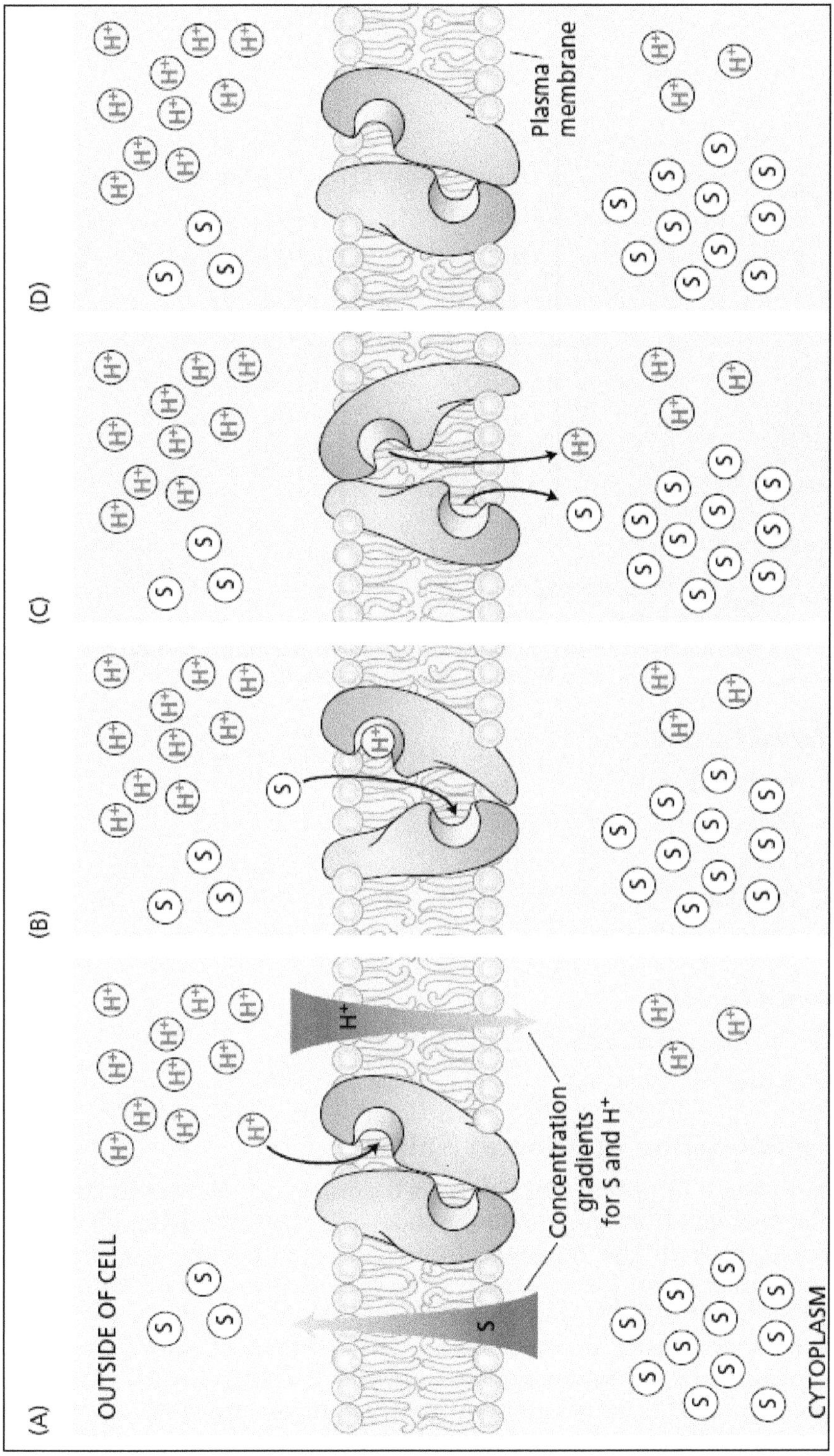

Figure 2.3: Secondary Active Transport Model.

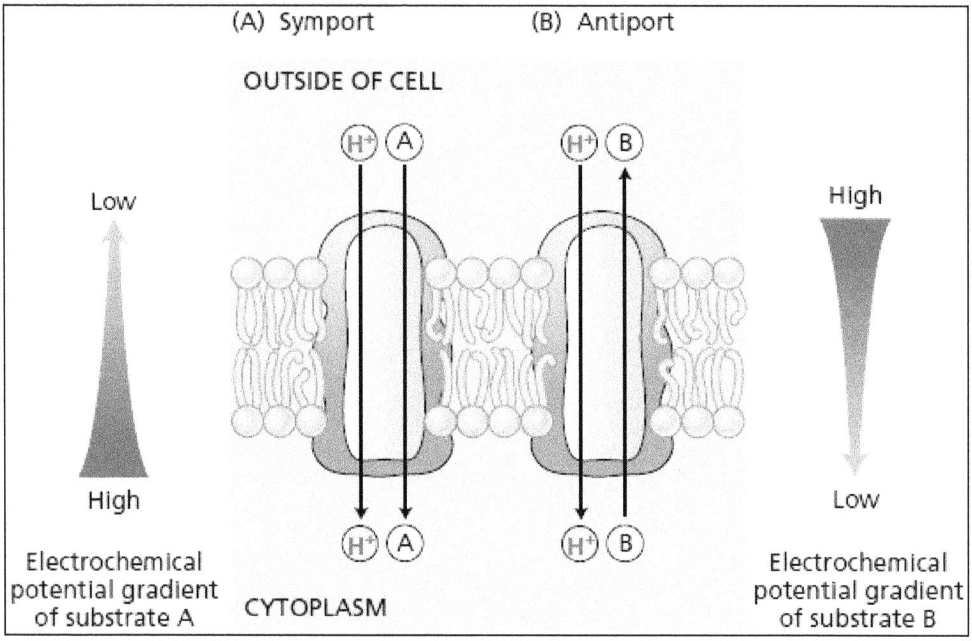

Figure 2.4: Example of Secondary Active Transport Coupled to a Primary Proton Gradient.

(A) External Factors

1. Temperature
2. $[H^+]$
3. Light
4. Oxygen
5. Interaction

(B) Internal Factors

1. Growth and morphophysiological status
2. Aging

2.5. Translocation of Mineral Salts

Both xylem and phloem take part in the transport of mineral salts. While upward movement, from the root to the shoot, mainly occurs through xylem, the bidirectional (upward and downward) movement, usually from mature leaves, occure through phloem. The transport of minerals in the xylem occurs along with the transpiration stream.This has been demonstrated by ringing experiments that upward transport of salts continues even when the phloem tissues were removed. Xylem sap also show the presence of large amount of dissolved salts. These include not only inorganic salts but also organic compounds of nitrogen and phosphorus.

Some of the minerals are transferred radially from xylem to the phloem mainly through parenchymatous cells. (Lateral transport).

2.5.1. Mechanism of Translocation

The ions are passively translocted through the wet cell walls and plasmodesmata of the cortex cells to the endodermis. Finally the ions are released in to xylem from the living cells located inward from endodermis.This is called symplast theory of ions uptake. Once the minerals enter in to xylem element, they are translocated upward along the transpiration stream.

References

Blevins, D.G. and Lukaszewski, K.M. (1999). Boron and plant struxture and function. *Annual Review of Plant Biology*, 49: 481-500.

Broadley, M.R. and White, P.J. (2005). *PlantNutritional Genomics*. Oxford, U.K., Blackwell Publishing.

Epstein, E.A. and Bloom, J. (2005). *Mineral Nutrition of Plants: Principles and Perspectives*. Sunderland: Sinauer Associates Inc.

Schachtman, D.P. and Shin, R. (2007). Nutritient sensing: NPKS. *Annual Review of Plant Biology*.

Chapter 3

Nitrogen Metabolism

In this chapter we will discuss the biogeochemical cycle of nitrogen, the crucial role of nitrogen fixation in the conversion of molecular nitrogen into ammonium and nitrate, and the fate of nitrate and ammonium in plant tissues.

3.1. Nitrogen Passes through Several Forms in a Biogeochemical Cycle

Nitrogen is present in many forms in the biosphere. The atmosphere contains vast quantities (about 78 per cent by volume) of molecular nitrogen (N_2). For the most part, this large reservoir of nitrogen is not directly available to living organisms. Acquisition of nitrogen from the atmosphere requires the breaking of an exceptionally stable triple covalent bond between two nitrogen atoms (N– N) to produce ammonia (NH_3) or nitrate (NO_3^-). These reactions, known as nitrogen fixation, can be accomplished by both industrial and natural processes. Natural processes fix the nitrogen (Table 3.1) through the following processes – lightning(about 8 per cent), photochemical reaction (about 2 per cent) and biological nitrogen fixation (about 90 per cent in which bacteria or blue-green algae (cyanobacteria) fix N_2 into ammonium NH_4^+).Once fixed in ammonium or nitrate, nitrogen enters a biogeochemical cycle and passes through several organic or inorganic forms before it eventually returns to molecular nitrogen.

3.2. Biological Nitrogen Fixation

The fixation of atmospheric N_2 into ammonium with the help of nitrogenase enzymes that fix nitrogen, the symbiotic relations between nitrogen-fixing organisms and higher plants.

3.3. Free-Living and Symbiotic Bacteria Fix Nitrogen

Some bacteria, as stated earlier, can convert atmospheric nitrogen into ammonium (Table 3.1). Most of these nitrogen- fixing prokaryotes are free-living in the soil. A few form symbiotic associations with higher plants in which the

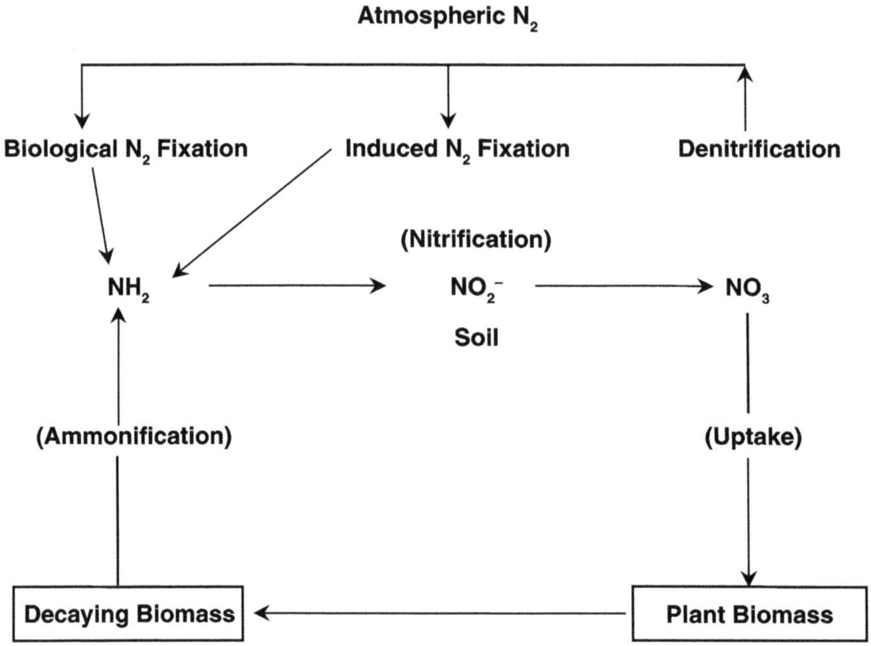

Figure 3.1: Nitrogen Cycle.

prokaryote directly provides the host plant with fixed nitrogen in exchange for other nutrients and carbohydrates (top portion of Table 3.1). Such symbioses occur in nodules that form on the roots of the plant and contain the nitrogen-fixing bacteria. The most common type of symbiosis occurs between members of the plant family Leguminosae and soil bacteria of the genera *Azorhizobium, Bradyrhizobium, Photorhizobium, Rhizobium,* and *Sinorhizobium* collectively called rhizobia (Table 4.1). Another common type of symbiosis occurs between several woody plant species, such as alder trees, and soil bacteria of the genus *Frankia*. Still other types involve the South American herb *Gunnera* and the tiny water fern *Azolla,* which form associations with the cyanobacteria *Nostoc* and *Anabaena,* respectively (Table 4.1).

3.4. Nitrogen Fixation Requires Anaerobic Conditions

Because oxygen irreversibly inactivates the nitrogenase enzymes involved in nitrogen fixation, nitrogen must be fixed under anaerobic conditions. Thus each of the nitrogen-fixing organisms listed in Table 3.1 either functions under natural anaerobic conditions or can create an internal anaerobic environment in the presence of oxygen. In cyanobacteria, anaerobic conditions are created in specialized cells called *heterocysts*. Heterocysts are thick-walled cells that differentiate when filamentous cyanobacteria are deprived of NH_4^+. These cells lack photosystem II, the oxygen-producing photosystem of chloroplasts, so they do not generate oxygen (Burris 1976). Heterocysts appear to represent an adaptation for nitrogen fixation, in that they are widespread among aerobic cyanobacteria that fix nitrogen. Cyanobacteria that lack heterocysts can fix nitrogen only under anaerobic

Table 3.1: Examples of Organisms that can carry out Nitrogen Fixation

Symbiotic nitrogen fixation	
Host plant	**N-fixing symbionts**
Leguminous: legumes, *Parasponia*	*Azorhizobium, Bradyrhizobium, Photorhizobium, Rhizobium, Sinorhizobium*
Actinorhizal: alder (tree), *Ceanothus* (shrub), *Casuarina* (tree), *Datisca* (shrub)	*Frankia*
Gunnera	*Nostoc*
Azolla (water fern)	*Anabaena*
Sugarcane	*Acetobacter*
Free-living nitrogen fixation	
Type	**N-fixing genera**
Cyanobacteria (blue-green algae)	*Anabaena, Calothrix, Nostoc*
Other bacteria	
Aerobic	*Azospirillum, Azotobacter, Beijerinckia, Derxia*
Facultative	*Bacillus, Klebsiella*
Anaerobic	
Nonphotosynthetic	*Clostridium, Methanococcus* (archaebacterium)
Photosynthetic	*Chromatium, Rhodospirillum*

conditions such as those that occur in flooded fields. In Asian countries, nitrogen-fixing cyanobacteria of both the heterocyst and nonheterocyst types are a major means for maintaining an adequate nitrogen supply in the soil of rice fields. These microorganisms fix nitrogen when the fields are flooded and die as the fields dry, releasing the fixed nitrogen to the soil.

Another important source of available nitrogen in flooded rice fields is the water fern *Azolla*, which associates with the cyanobacterium *Anabaena*. The *Azolla–Anabaena* association can fix as much as 0.5 kg of atmospheric nitrogen per hectare per day, a rate of fertilization that is sufficient to attain moderate rice yields. Free-living bacteria that are capable of fixing nitrogen are aerobic, facultative, or anaerobic (Table 3.1):

Table 3.2: Associated between Host Plants and Rhizobia

Plant host	Rhizobial symbiont
Parasponia (a nonlegume, formerly called *Trema*)	*Bradyrhizobium* spp.
Soybean (*Glycine max*)	*Bradyrhizobium japonicum* (slow-growing type); *Sinorhizobium fredii* (fast-growing type)
Alfalfa (*Medicago sativa*)	*Sinorhizobium meliloti*
Sesbania (aquatic)	*Azorhizobium* (forms both root and stem nodules; the stems have adventitious roots)
Bean (*Phaseolus*)	*Rhizobium leguminosarum* bv. *phaseoli*; *Rhizobium tropicii*; *Rhizobium etli*
Clover (*Trifolium*)	*Rhizobium leguminosarum* bv. *trifolii*
Pea (*Pisum sativum*)	*Rhizobium leguminosarum* bv. *viciae*
Aeschenomene (aquatic)	*Photorhizobium* (photosynthetically active rhizobia that form stem nodules, probably associated with adventitious roots)

- *Aerobic* nitrogen-fixing bacteria such as *Azotobacter* are thought to maintain reduced oxygen conditions (microaerobic conditions) through their high levels of respiration (Burris 1976). Others, such as *Gloeothece*, evolve O2 photosynthetically during the day and fix nitrogen during the night.
- *Facultative* organisms, which are able to grow under both aerobic and anaerobic conditions, generally fix nitrogen only under anaerobic conditions.
- For *anaerobic* nitrogen-fixing bacteria, oxygen does not pose a problem, because it is absent in their habitat. These anaerobic organisms can be either photosynthetic (*e.g., Rhodospirillum*), or nonphotosynthetic (*e.g., Clostridium*).

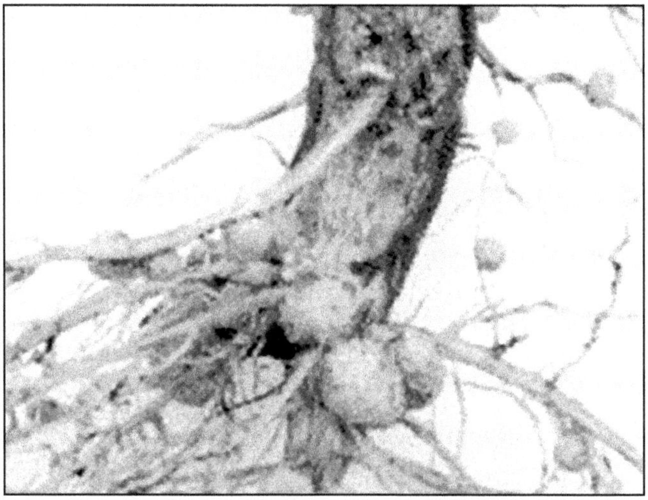

Figure 3.1: Symbiotic Nitrogen Fixing Prokaryotes.

Symbiotic nitrogen-fixing prokaryotes dwell within nodules, the special organs of the plant host that enclose the nitrogen-fixing bacteria (Figure 3.1). In the case of legumes and actinorhizal plants, the nitrogen-fixing bacteria induce the plant to form root nodules. Grasses can also develop symbiotic relationships with nitrogen-fixing organisms, but in these associations root nodules are not produced. Instead, the nitrogen-fixing bacteria seem to colonize plant tissues or anchor to the root surfaces, mainly around the elongation zone and the root hairs (Reis *et al.*, 2000). Legumes and actinorhizal plants regulate gas permeability in their nodules, maintaining a level of oxygen within the nodule that can support respiration but is sufficiently low to avoid inactivation of the nitrogenase (Kuzma *et al.*, 1993). Nodules contain an oxygen-binding heme protein called leghemoglobin. Leghemoglobin is present in the cytoplasm of infected nodule cells at high concentrations and gives the nodules a pink color. The host plant produces the globin portion of leghemoglobin in response to infection by the bacteria (Marschner 1995); the bacterial symbiont produces the heme portion.Leghemoglobin has a high affinity for oxygen, about ten times higher than the ??chain of human hemoglobin.

3.5. Nodulation and Nod Factors

Plant genes specific to nodules are called nodulin (*Nod*) genes; rhizobial genes that participate in nodule formation are called nodulation (*nod*) genes (Heidstra and Bisseling 1996). The *nod* genes are classified as common *nod* genes or host-specific *nod* genes. The common *nod* genes–*nodA*, *nodB*, and *nodC*–are found in all rhizobial strains; the host-specific *nod* genes–such as *nodP*, *nodQ*, and *nodH*; or *nodF*, *nodE*, and *nodL*–differ among rhizobial species and determine the host range. Only one of the *nod* genes, the regulatory *nodD*, is constitutively expressed, and as we will explain in detail, its protein product (NodD) regulates the transcription of the other *nod* genes. The first stage in the formation of the symbiotic relationship between the nitrogen-fixing bacteria and their host is migration of the bacteria toward the roots of the host plant. This migration is a chemotactic response mediated by chemical attractants, especially (iso)flavonoids and betaines, secreted by the roots. These attractants activate the rhizobial NodD protein, which then induces transcription of the other *nod* genes (Phillips and Kapulnik 1995). The promoter region of all *nod* operons, except that of *nodD*, contains a highly conserved sequence called the *nod* box. Binding of the activated NodD to the *nod* box induces transcription of the other *nod* genes.

The *nod* genes activated by Nod D code for nodulation proteins, most of which are involved in the biosynthesis of Nod factors. Nod factors are lipochitin oligosaccharide signal molecules, all of which have a chitin α-1$\rightarrow$4-linked *N*acetyl-D-glucosamine backbone (varying in length from three to six sugar units) and a fatty acyl chain on the C-2 position of the nonreducing sugar. Three of the *nod* genes (*nodA*, *nodB*, and *nodC*) encode enzymes (NodA, NodB, and NodC, respectively) that are required for synthesizing this basic structure.

1. NodA is an *N*-acyltransferase that catalyzes the addition of a fatty acyl chain.
2. NodB is a chitin-oligosaccharide deacetylase that removes the acetyl group from the terminal nonreducing sugar.
3. NodC is a chitin-oligosaccharide synthase that links *N*-acetyl-D-glucosamine monomers.

Host-specific *nod* genes that vary among rhizobial species are involved in the modification of the fatty acyl chain or the addition of groups important in determining host specificity (Carlson *et al.*, 1995):

- ☆ NodE and NodF determine the length and degree of saturation of the fatty acyl chain; those of *Rhizobium leguminosarum* bv. *viciae* and *R. meliloti* result in the synthesis of an 18:4 and a 16:2 fatty acyl group, respectively.
- ☆ Other enzymes, such as NodL, influence the host specificity of Nod factors through the addition of specific substitutions at the reducing or nonreducing sugar moieties of the chitin backbone. Aparticular legume host responds to a specific Nod factor. The legume receptors for Nod factors appear to be special lectins (sugar-binding proteins) produced in the root hairs. Nod factors activate these lectins, increasing their hydrolysis

of phosphoanhydride bonds of nucleoside di- and triphosphates. This lectin activation directs particular rhizobia to appropriate hosts and facilitate attachment of the rhizobia to the cell walls of a root hair.

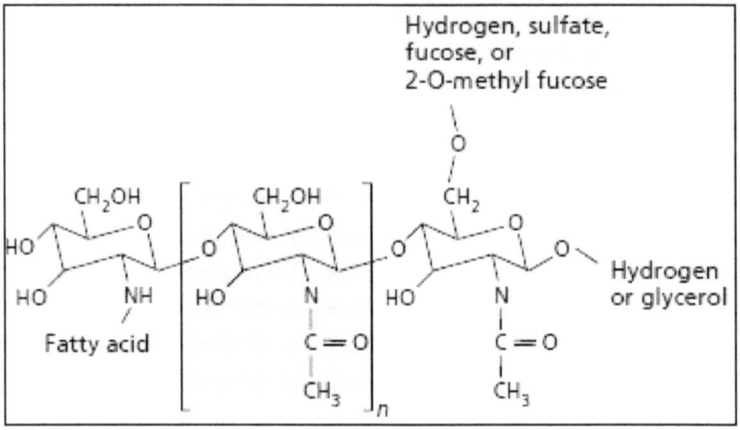

Figure 3.2

3.5.1. Nodule Formation

Two processes–infection and nodule organogenesis– occur simultaneously during root nodule formation. During the infection process, rhizobia that are attached to the root hairs release Nod factors that induce a pronounced curling of the root hair cells. The rhizobia become enclosed in the small compartment formed by the curling. The cell wall of the root hair degrades in these regions, also in response to Nod factors, allowing the bacterial cells direct access to the outer surface of the plant plasma membrane (Lazarowitz and Bisseling 1997). The next step is formation of the infection thread (Figure 3.3) an internal tubular extension of the plasma membrane that is produced by the fusion of Golgi-derived membrane vesicles at the site of infection. The thread grows at its tip by the fusion of secretory vesicles to the end of the tube. Deeper into the root cortex, near the xylem, cortical cells dedifferentiate and start dividing, forming a distinct area within the cortex, called a *nodule primordium*, from which the nodule will develop. The nodule primordia form opposite the protoxylem poles of the root vascular bundle.Different signaling compounds, acting either positively or negatively, control the position of nodule primordia. The nucleoside uridine diffuses from the stele into the cortex inthe protoxylem zones of the root and stimulates cell division (Lazarowitz and Bisseling 1997). Ethylene is synthesized in the region of the pericycle, diffuses into the cortex, and blocks cell division opposite the phloem poles of the root. The infection thread filled with proliferating rhizobia elongates through the root hair and cortical cell layers, in the direction of the nodule primordium. When the infection thread reaches specialized cells within the nodule, its tip fuses with the plasma membrane of the host cell, releasing bacterial cells that are packaged in a membrane derived from the host cell plasma membrane (Figure 3.3 D). Branching of the infection thread inside the nodule enables the bacteria to infect many cells (Figure 3.3 E and F)

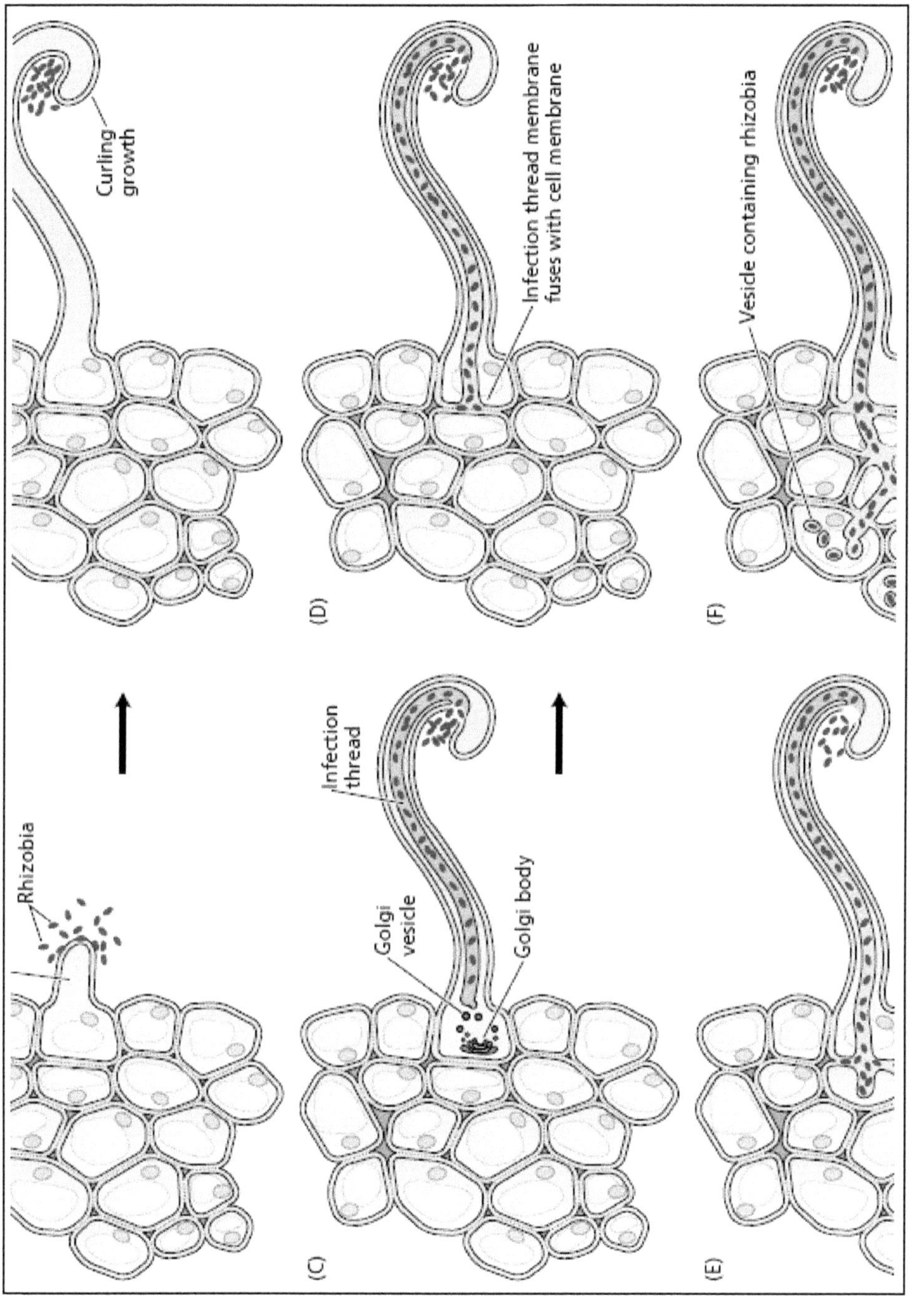

Figure 3.3: Steps of Nodule Formation.

(Mylona *et al.*, 1995). At first the bacteria continue to divide, and the surrounding membrane increases in surface area to accommodate this growth by fusing with smaller vesicles. Soon thereafter, upon an undetermined signal from the plant, the bacteria stop dividing and begin to enlarge and to differentiate into nitrogen-fixing endosymbiotic organelles called bacteroids. The membrane surrounding the bacteroids is called the *peribacteroid membrane*. The nodule as a whole develops such features as a vascular system (which facilitates the exchange of fixed nitrogen produced by the bacteroids for nutrients contributed by the plant) and a layer of cells to exclude O2 from the root nodule interior. In some temperate legumes (*e.g.*, peas), the nodules are elongated and cylindrical because of the presence of a *nodule meristem*. The nodules of tropical legumes, such as soybeans and peanuts, lack a persistent meristem and are spherical.

3.5.2. The Nitrogenase Enzyme Complex Fixes N_2

Biological nitrogen fixation, like industrial nitrogen fixation, produces ammonia from molecular nitrogen. The overall reaction is

$$N_2 + 8\ e- + 8\ H+ + 16\ ATP \rightarrow 2\ NH_3 + H_2 + 16\ ADP + 16\ Pi$$

The reduction of N2 to 2 NH3, a six-electrontransfer, is coupled to the reduction of two protons to evolve H_2. The nitrogenase enzyme complex catalyzes this reaction. The nitrogenase enzyme complex can be separated into two components–the Fe protein and the MoFe protein– neither of which has catalytic activity by itself (Figure 3.4):

- ✰ The Fe protein is the smaller of the two components and has two identical subunits of 30 to 72 kDa each, depending on the organism. Each subunit contains an iron–sulfur cluster (4 Fe and 4 S_2–) that participates in the redox reactions involved in the conversion of N_2 to NH_3. The Fe protein is irreversibly inactivated by O_2 with typical half-decay times of 30 to 45 seconds.
- ✰ The MoFe protein has four subunits, with a total molecular mass of 180 to 235 kDa, depending on the species. Each subunit has two Mo–Fe–S clusters. The MoFe protein is also inactivated by oxygen, with a half-decay time in air of 10 minutes.

In the overall nitrogen reduction reaction (Figure 3.4), ferredoxin serves as an electron donor to the Fe protein, which in turn hydrolyzes ATP and reduces the MoFe protein. The MoFe protein then can reduce numerous substrates (Table 3.3), although under natural conditions it reacts only with N_2 and H^+. One of the reactions catalyzed by nitrogenase, the reduction of acetylene to ethylene, is used in estimating nitrogenase activity. The energetics of nitrogen fixation is complex. The production of NH_3 from N_2 and H_2 is an exergonic reaction. However, industrial production of NH_3 from N_2 and H_2 is *endergonic*, requiring a very large energy input because of the activation energy needed to break the triple bond in N_2. For the same reason, the enzymatic reduction of N_2 by nitrogenase also requires a large investment of energy (see Equation), the overall reaction is highly exergonic, ammonium production is limited by the slow operation (number of N_2 molecules reduced per unit time)

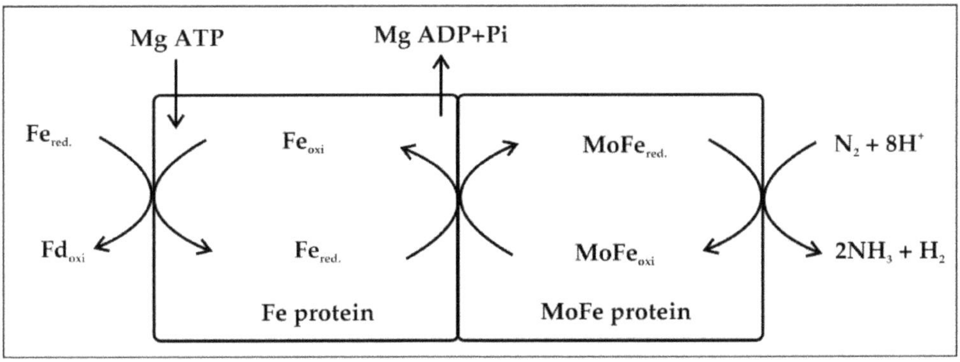

Figure 3.4: Nitrogenase Enzyme Complex.

the nitrogenase complex (Ludwig and de Vries 1986). Under natural conditions, substantial amounts of H^+ are reduced to H_2 gas, and this process can compete with N_2 reduction for electrons from nitrogenase. In rhizobia, 30 to 60 per cent of the energy supplied to nitrogenase may be lost as H_2, diminishing the efficiency of nitrogen fixation. Some rhizobia, however, contain hydrogenase, an enzyme that can split the H_2 formed and generate electrons for N_2 reduction, thus improving the efficiency of nitrogen fixation.

Table 3.3: Reactions Catalyzed by Nitrogenase

$N_2 \rightarrow NH_3$	Molecular nitrogen fixation
$N_2O \rightarrow N_2 + H_2O$	Nitrous oxide reduction
$N_3^- \rightarrow N_2 + NH_3$	Azide reduction
$C_2H_2 \rightarrow C_2H_4$	Acetylene reduction
$2H^+ \rightarrow H_2$	H_2 production
ATP → ADP + Pi	ATP hydrolytic activity

3.6. Nitrate Assimilation

Plants assimilate most of the nitrate absorbed by their roots into organic nitrogen compounds. The first step of this process is the reduction of nitrate to nitrite in the cytosol (Oaks 1994). The enzyme nitrate reductase catalyzes this reaction:

$$NO_3^- + NAD(P)H + H^+ + 2\,e^- \rightarrow NO_2^- + NAD(P)^+ + H_2O$$

where, NAD(P)H indicates NADH or NADPH. The most common form of nitrate reductase uses only NADH as an electron donor; another form of the enzyme that is found predominantly in nongreen tissues such as roots can use either NADH or NADPH (Warner and Kleinhofs, 1992). The nitrate reductases of higher plants are composed of two identical subunits, each containing three prosthetic groups: FAD (flavin adenine dinucleotide), heme, and a molybdenum complexed to an organic molecule called a *pterin*. Nitrate reductase is the main molybdenum-containing protein in vegetative tissues, and one symptom of molybdenum deficiency is the accumulation of nitrate that results from diminished nitrate reductase activity.

Comparison of the amino acid sequences for nitrate reductase from several species with those of other well characterized proteins that bind FAD, heme, or molybdenum has led to the three-domain model for nitrate reductase. The FAD-binding domain accepts two electrons from NADH or NADPH. The electrons then pass through the heme domain to the molybdenum complex, where they are transferred to nitrate.

Regulation of nitrate reductase activity through phosphorylation and dephosphorylation provides more rapid control than can be achieved through synthesis or degradation of the enzyme.

3.7. Nitrite Reductase Converts Nitrite to Ammonium

Nitrite (NO_2^-) is a highly reactive, potentially toxic ion. Plant cells immediately transport the nitrite generated by nitrate reduction (see Equation) from the cytosol into chloroplasts in leaves and plastids in roots. In these organelles, the enzyme nitrite reductase reduces nitrite to ammonium according to the following overall reaction:

$$NO_2^- + 6\ Fdred + 8\ H^+ + 6\ e^- \rightarrow NH_4^+ + 6\ Fdox + 2\ H_2O$$

where, Fd is ferredoxin, and the subscripts *red* and *ox* stand for *reduced* and *oxidized*, respectively. Reduced ferredoxin derives from photosynthetic electron transport in the chloroplasts and from NADPH generated by the oxidative pentose phosphate pathway in nongreen tissues. Chloroplasts and root plastids contain different forms of the enzyme, but both forms consist of a single polypeptide containing two prosthetic groups: an iron–sulfur cluster (Fe_4S_4) and a specialized heme (Siegel and Wilkerson 1989). These groups acting together bind nitrite and reduce it directly to ammonium, without accumulation of nitrogen compounds of intermediate redox states. The electron flow through ferredoxin (Fe_4S_4) and heme can be represented as in Figure 3.5. Nitrite reductase is encoded in the nucleus and synthesize in the cytoplasm with an N-terminal transit peptide that targets it to the plastids (Wray, 1993). Whereas NO_3^- and light induce the transcription of nitrite reductase mRNA, the end products of the process–asparagine and glutamine–repress this induction.

In many plants, when the roots receive small amounts of nitrate, nitrate is reduced primarily in the roots. As the supply of nitrate increases, a greater proportion of theabsorbed nitrate is translocated to the shoot and assimilated there.

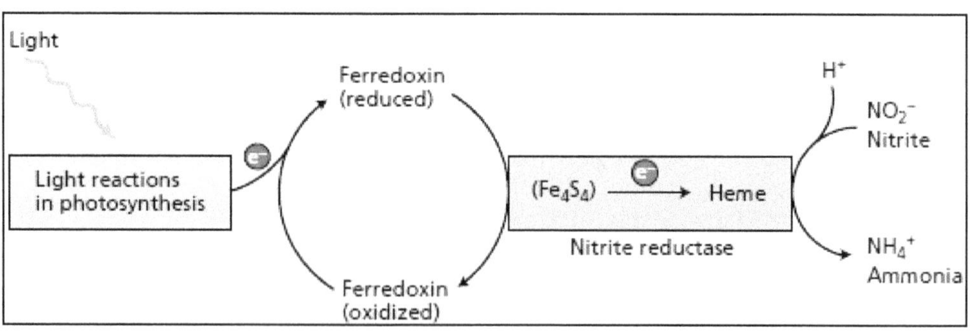

Figure 3.5: GS–GOGAT Pathways.

3.8. Ammonium Assimilation

Plant cells avoid ammonium toxicity by rapidly converting the ammonium generated from nitrate assimilation or photorespiration into amino acids. The primary pathway for this conversion involves the sequential actions of glutamine synthetase (GS) and glutamate synthase (GOGAT).

Conversion of Ammonium to Amino Acids

Requires Two Enzymes

Glutamine synthetase (GS) combines ammonium with glutamate to form glutamine (Figure 3.6):

$$\text{Glutamate} + NH_4^+ + \text{ATP} \rightarrow \text{glutamine} + \text{ADP} + \text{Pi}$$

This reaction requires the hydrolysis of one ATP and involves a divalent cation such as Mg_2^+, Mn_2^+, or CO_2^+ as a cofactor. Plants contain two classes of GS, one in the cytosol and the other in root plastids or shoot chloroplasts. The cytosolic forms are expressed in germinating seeds or in the vascular bundles of roots and shoots and produce glutamine for intracellular nitrogen transport. The GS in root plastids generates amide nitrogen for local consumption; the GS in shoot chloroplasts reassimilates photorespiratory NH_4^+ (Lam *et al.*, 1996). Light and carbohydrate levels alter the expression of the plastid forms of the enzyme, but they have little effect on the cytosolic forms. Elevated plastid levels of glutamine stimulate the activity of glutamate synthase (also known as *glutamine:2-oxoglutarate aminotransferase*, or GOGAT). This enzyme transfers the amide group of glutamine to 2-oxoglutarate, yielding two molecules of glutamate (Figure 3.6). Plants contain two types of GOGAT: One accepts electrons from NADH; the other accepts electrons from ferredoxin (Fd).

$$\text{Glutamine} + \text{2-oxoglutarate} + \text{NADH} + H^+ \rightarrow 2\ \text{glutamate} + \text{NAD}^+$$

$$\text{Glutamine} + \text{2-oxoglutarate} + \text{Fdred} \longrightarrow 2\ \text{glutamate} + \text{Fdox}$$

The NADH type of the enzyme (NADH-GOGAT) is located in plastids of nonphotosynthetic tissues such as roots or vascular bundles of developing leaves. In roots, NADH-GOGAT is involved in the assimilation of NH4+ absorbed from the rhizosphere (the soil near the surface of the roots); in vascular bundles of developing leaves, NADH-GOGAT assimilates glutamine translocated from roots or senescing leaves. The ferredoxin-dependent type of glutamate synthase (Fd-GOGAT) is found in chloroplasts and serves in photorespiratory nitrogen metabolism. Both the amount of protein and its activity increase with light levels. Roots, particularly those under nitrate nutrition, have Fd-GOGAT in plastids. Fd- GOGAT in the roots presumably functions to incorporate the glutamine generated during nitrate assimilation.

Ammonium can be Assimilated via an Alternative Pathway

Glutamate dehydrogenase (GDH) catalyzes a reversible reaction that synthesizes or deaminates glutamate.

$$\text{2-Oxoglutarate} + NH_4^+ + \text{NAD(P)H} \rightarrow \text{Glutamate} + H_2O + \text{NAD(P)}^+$$

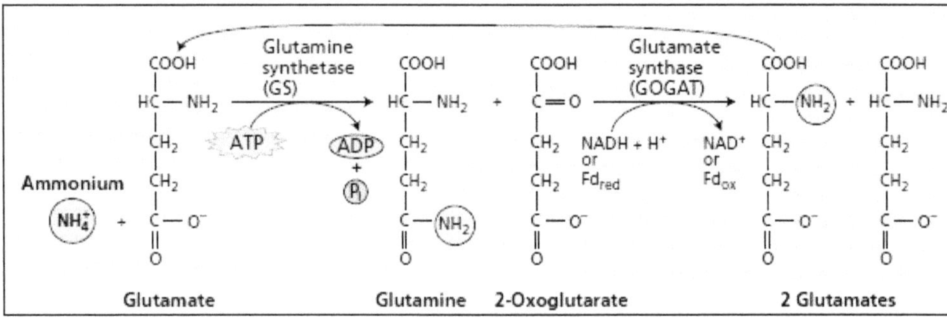

An NADH-dependent form of GDH is found in mitochondria, and an NADPH-dependent form is localized in the chloroplasts of photosynthetic organs. Although both forms are relatively abundant, they cannot substitute for the GS–GOGAT pathway for assimilation of ammonium, and their primary function is to deaminate glutamate.

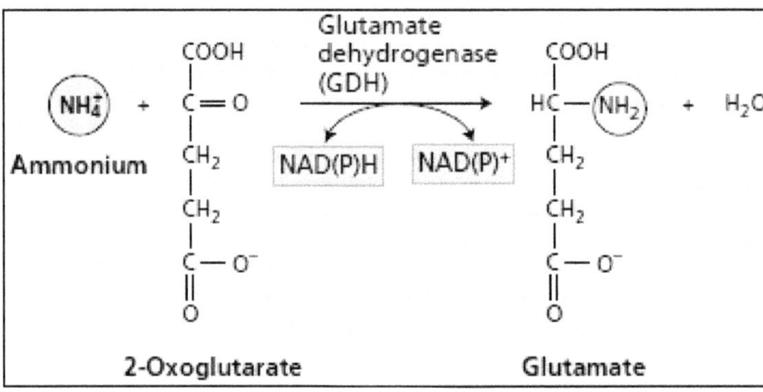

3.9. Transamination Reactions Transfer Nitrogen

Once assimilated into glutamine and glutamate, nitrogen is incorporated into other amino acids via transamination reactions. The enzymes that catalyze these reactions are known as aminotransferases. An example is aspartate aminotransferase (Asp-AT).

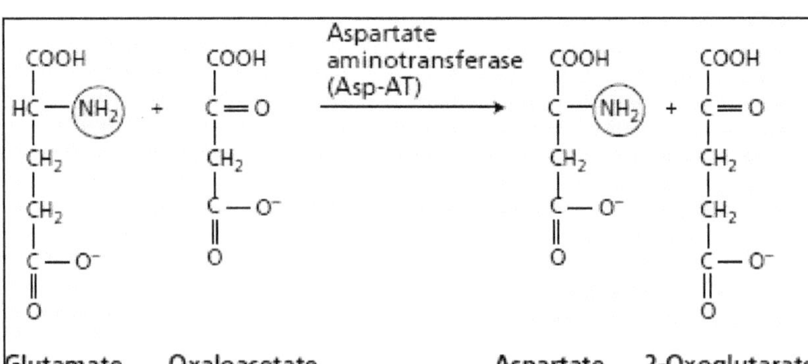

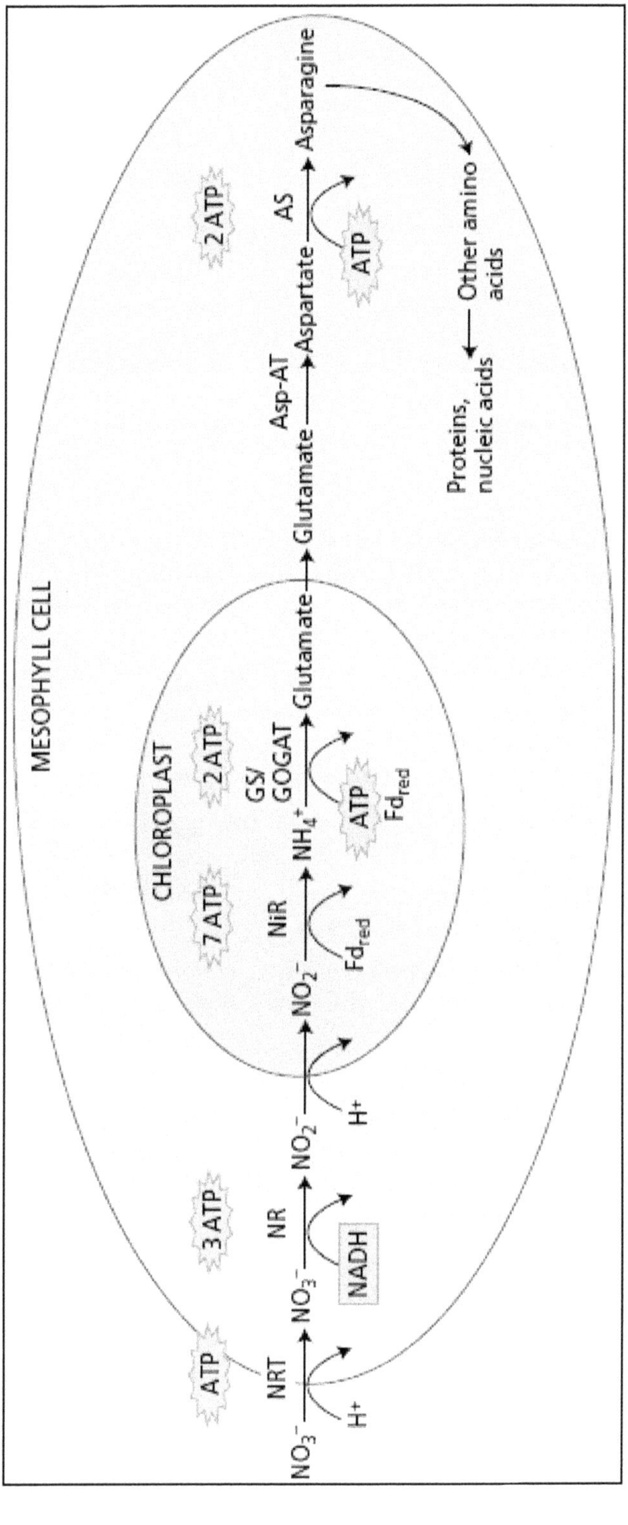

Figure 3.7: Summary of the Processes Involved in the Assimilation of Mineral Nitrogen in the Leaf.

The major pathway for asparagine synthesis involves the transfer of the amide nitrogen from glutamine to Asparagines.

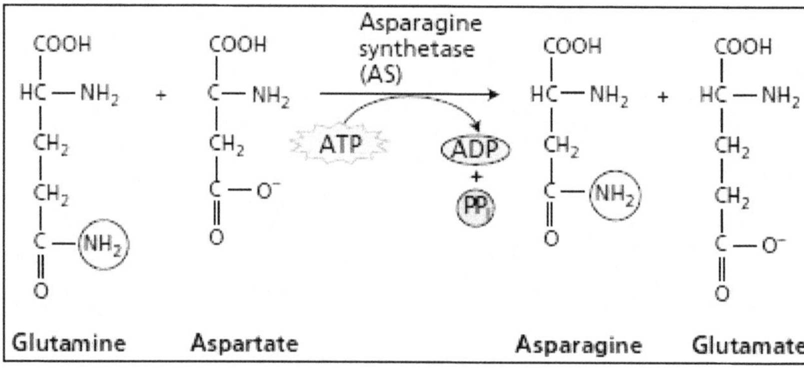

References

Andersen, P.C., B.V. Brodbeck and Herzog, D.C. (2002). Girdling-induced nutrient accumulation in aboveground tissue of peanuts and subsequent feeding by *Spissistilus festinus*, the three-cornered alfalfa hopper. *Entomology Experimental Application*. 103: 139–149.

Barnes, J.D., T. Pfirrmann, K. Steiner, C. Lutz, U. Busch, H. Kuchenhoff and Payer, H.D. (1995). Effects of elevated CO_2 elevated O_3 and potassium deficiency on Norway spruce [*Picea abies*(L.) Karst.]: seasonal changes in photosynthesis and nonstructural carbohydrate content. *Plant Cell Environ*. 18: 1345–1357.

Buchanan-Wollaston, V. (1997). The molecular biology of leaf senescence. *Journal of Experimental Botany*. 48: 181–199.

Chalker-Scott, L. (1999). Environmental significance of anthocyanins in plant stress responses. Photochem. *Photobiol*. 70:1–9.

Chang, K.G., G.H. Fechner and Schroeder, H.A. (1989). Anthocyanins in autumn leaves of quaking aspen in Colorado. *Forest. Science*. 35: 229–236.

Close, D.C. and Beadle, C.L. (2003). The ecophysiology of foliar anthocyanin. *Botanical Review*. 69:149–161.

Costa-Arbulú, C., E. Gianoli, W.L. Gonzáles and Niemeyer, H.M. (2001). Feeding by the aphid *Sipha flava* produces a reddish spot on leaves of *Sorghum halepense*: an induced defense? *Journal of Chemical Ecolology*. 27: 273–283.

Dube, A., S. Bharti and Laloraya. M.M. (1993). Inhibition of anthocyanin synthesis and phenylalanine ammonialyase activity by CO_2^+ in leaf disks of *Terminalia catappa*. *Physiology Plant*. 88:237–242.

Einig,W., U. Lauxmann, B. Hauch, R. Hampp,W. Landolt, S. Maurer and R. Matyssek. 1997. Ozone-induced accumulation of carbohydrates changes enzyme activities of carbohydrate metabolism in birch leaves. *New Phytology*. 137:673–680.

Field, T.S., D.W. Lee and N.M. Holbrook.(2001). Why leaves turn red in autumn. The role of anthocyanins in senescing leaves of redosier dogwood. *Plant Physiology*. 127:566–574.

Chapter 4

Photosynthesis

Life on earth ultimately depends on energy derived from the sun. The term *photosynthesis* means literally "synthesis using light." As we will see in this chapter, photosynthetic organisms use solar energy to synthesize carbon compounds that cannot be formed without the input of energy. More specifically, light energy drives the synthesis of carbohydrates from carbon dioxide and water with the generation of oxygen:

$$6\,CO_2 + 12\,H_2O \rightarrow C_6H_{12}O_6 + 6\,O_2 + 6H_2O$$

Carbon dioxide   Water   Carbohydrate   Oxygen

Energy stored in these molecules can be used later to power cellular processes in the plant and can serve as the energy source for all forms of life.

The most active photosynthetic tissue in higher plants is the mesophyll of leaves. Mesophyll cells have many chloroplasts, which contain the specialized light-absorbing green pigments, the chlorophylls. In photosynthesis, the plant uses solar energy to oxidize water, thereby releasing oxygen, and to reduce carbon dioxide, thereby forming large carbon compounds, primarily sugars. The complex series of reactions that culminate in the reduction of CO_2 include the thylakoid reactions and the carbon fixation reactions.The thylakoid reactions of photosynthesis take place in the specialized internal membranes of the chloroplast called thylakoids. The end products of these thylakoid reactions are the high-energy compounds ATP and NADPH, which are used for the synthesis of sugars in the carbon fixation reactions. These synthetic processes take place in the stroma of the chloroplasts.

4.1. Photosynthetic Apparatus

Chloroplast

In photosynthetic eukaryotes, photosynthesis takes place in the subcellular organelle known as the chloroplast. The most striking aspect of the structure of the chloroplast is the extensive system of internal membranes known as thylakoids. All the chlorophyll is contained within this membrane system, which is the site of the light reactions of photosynthesis. The carbon reduction reactions, which are catalyzed by water-soluble enzymes, take place in the stroma (plural *stromata*), the region of the chloroplast outside the thylakoids. Most of the thylakoids appear to be very closely associated with each other. These stacked membranes are known as grana lamellae (singular *lamella*; each stack is called a *granum*), and the exposed membranes in which stacking is absent are known as stroma lamellae. Two separate membranes, each composed of a lipid bilayer and together known as the envelope, surround most types of chloroplasts (Figure 4.1). This double-membrane system contains a variety of metabolite transport systems.

The chloroplast also contains its own DNA, RNA, and ribosomes. Many of the chloroplast proteins are products of transcription and translation within the chloroplast itself, whereas others are encoded by nuclear DNA, synthesized on cytoplasmic ribosomes, and then imported into the chloroplast. This remarkable division of labor, extending in many cases to different subunits of the same enzyme

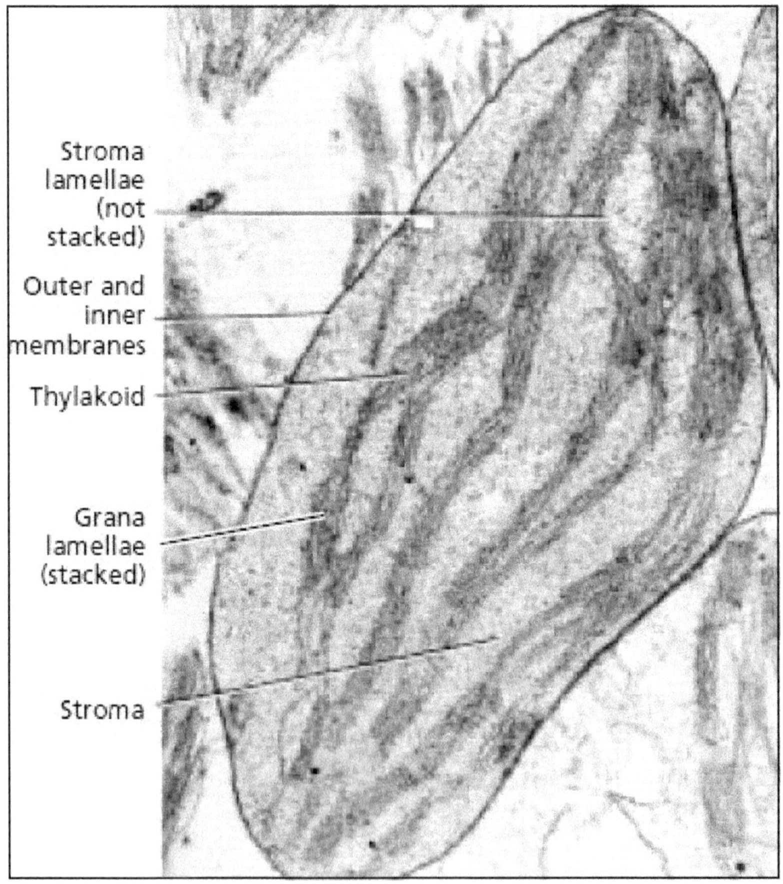

Figure 4.1a: Transmission Electron Micrograph of Chloroplast from Pea (Courtesy of Taiz and Zieger).

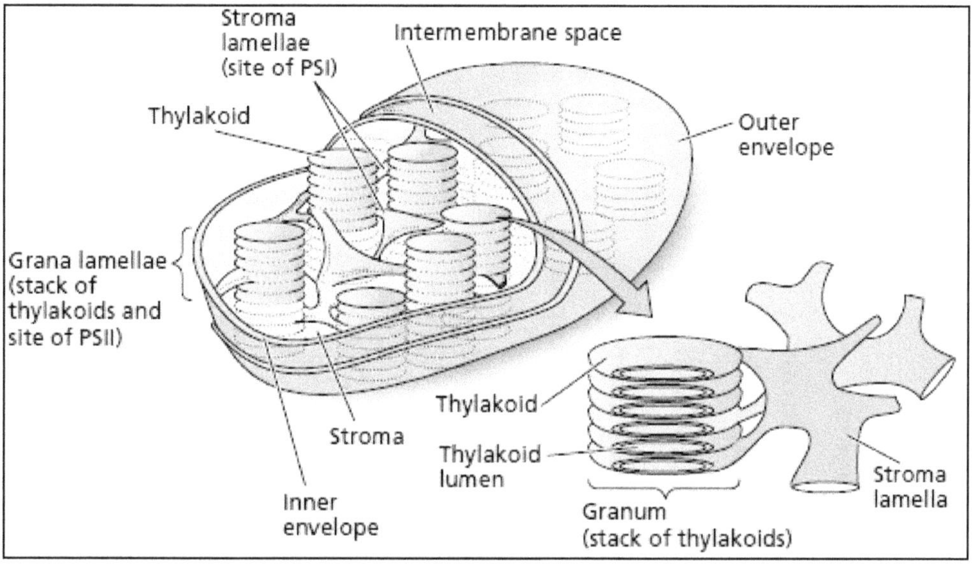

Figure 4.1b: Organization of the Membranes in the Chloroplast (Courtsey of Taiz and Zieger).

complex such as larger subunit (rbcL.) of Rubisco encoded by chloroplast gene while smaller subunit (rbcS) of Rubisco encoded by nuclear gene. But in brown and red algae, both subunit of Rubisco encoded by chloroplast gene.

A wide variety of proteins essential to photosynthesis are embedded in the thylakoid membranes. In many cases, portions of these proteins extend into the aqueous regions on both sides of the thylakoids. These integral membrane proteins contain a large proportion of hydrophobic amino acids and are therefore much more stable in a non aqueous medium such as the hydrocarbon portion of the membrane. The reaction centers, the antenna pigment– protein complexes, and most of the electron transport enzymes are all integral membrane proteins. In all known cases, integral membrane proteins of the chloroplast have a unique orientation within the membrane. Thylakoid membrane proteins have one region pointing toward the stromal side of the membrane and the other oriented toward the interior portion of the thylakoid, known as the *lumen.* The chlorophylls and accessory light-gathering pigments in the thylakoid membrane are always associated in a noncovalent but highly specific way with proteins. Both antenna and reaction center chlorophylls are associated with proteins that are organized within the membrane so as to optimize energy transfer in antenna complexes and electron transfer in reaction centers, while at the same time minimizing wasteful processes.

Thylakoid membrane is composed of lipo-proteinic subunits. The protein core coated on the outside with lipids. Chlorophyll, carotenoid molecules are concentrated in the appressed region.The head of chlorophyll molecules occupy the area between the subunits while phytol tail are buried in the interior of the subunits. All chlorophyll molecule are non covalently associated with protein.

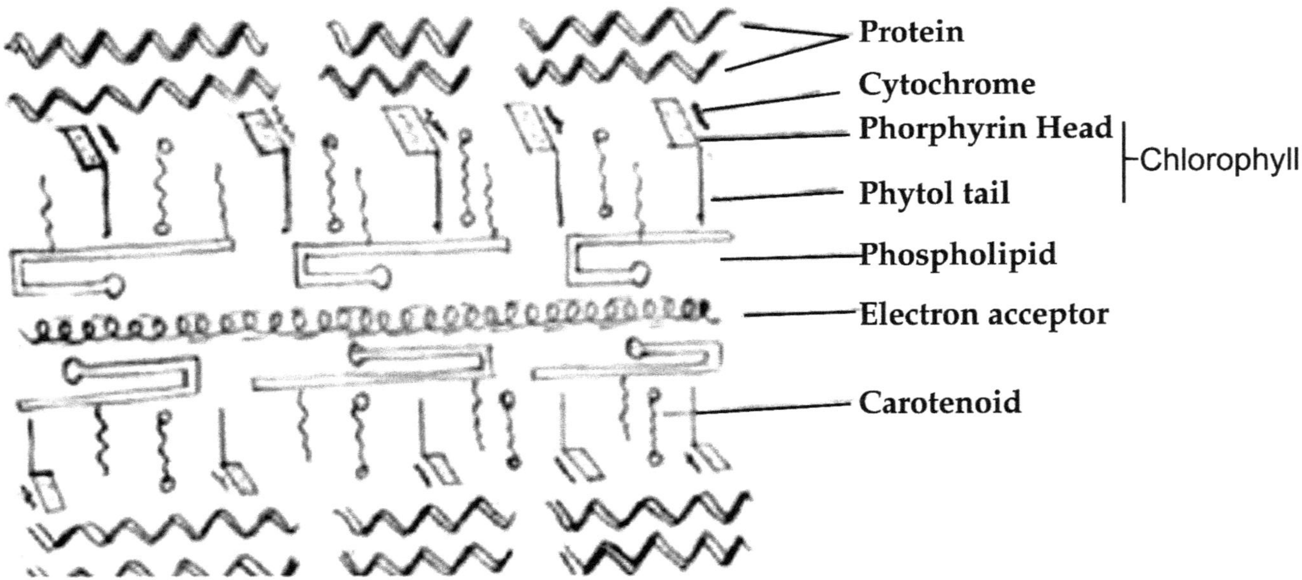

Figure 4.2: The Protein is Asymmetrically Arranged in the Thylakoid Membrane (Courtesy of Taiz and Zieger).

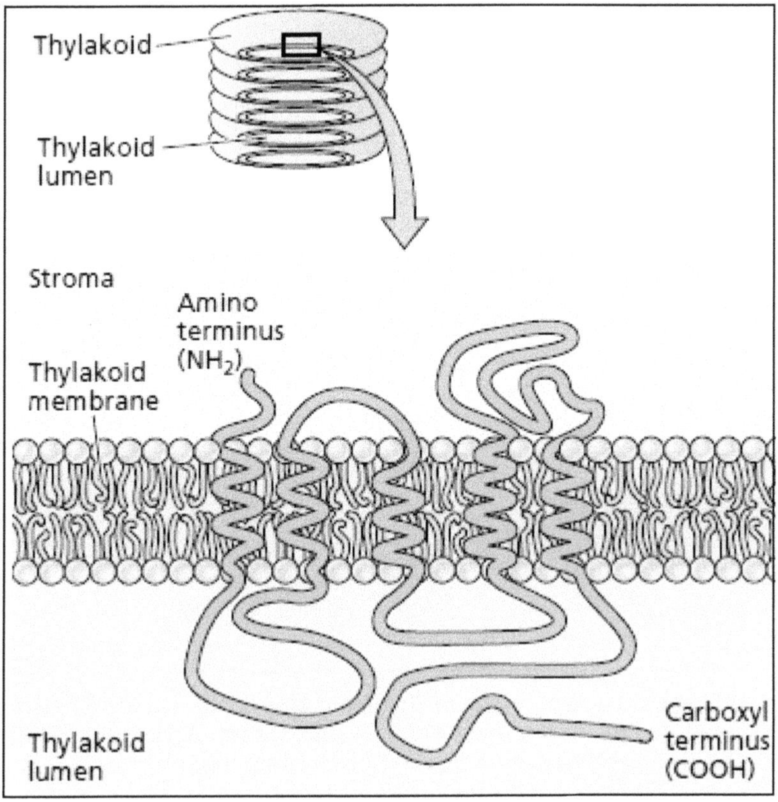

Figure 4.3: Molecular Arrangement of Thylakoid Membrane.

The PSII reaction center, along with its antenna chlorophylls and associated electron transport proteins, is located predominantly in the grana lamellae (Figure 4.4) (Allen and Forsberg, 2001). The PSI reaction center and its associated antenna pigments and electron transfer proteins, as well as the coupling- factor enzyme that catalyzes the formation of ATP, are found almost exclusively in the stroma lamellae and at the edges of the grana lamellae. The cytochrome *b*6 *f* complex of the electron transport chain that connects the two photosystems is evenly distributed between stroma and grana. Thus the two photochemical events that take place in O_2-evolving photosynthesis are spatially separated. This separation implies that one or more of the electron carriers that function between the photosystems diffuses from the grana region of the membrane to the stroma region, where electrons are delivered to photosystem I. In PSII, the oxidation of two water molecules produces four electrons, four protons, and a single O_2. The protons produced by this oxidation of water must also be able to diffuse to the stroma region, where ATP is synthesized. In addition to the vectorial arrangement of electron transport components across the thylakoid membrane, there is also a distinct lateral heterogeneity with respect to their distribution within the thylakoids.

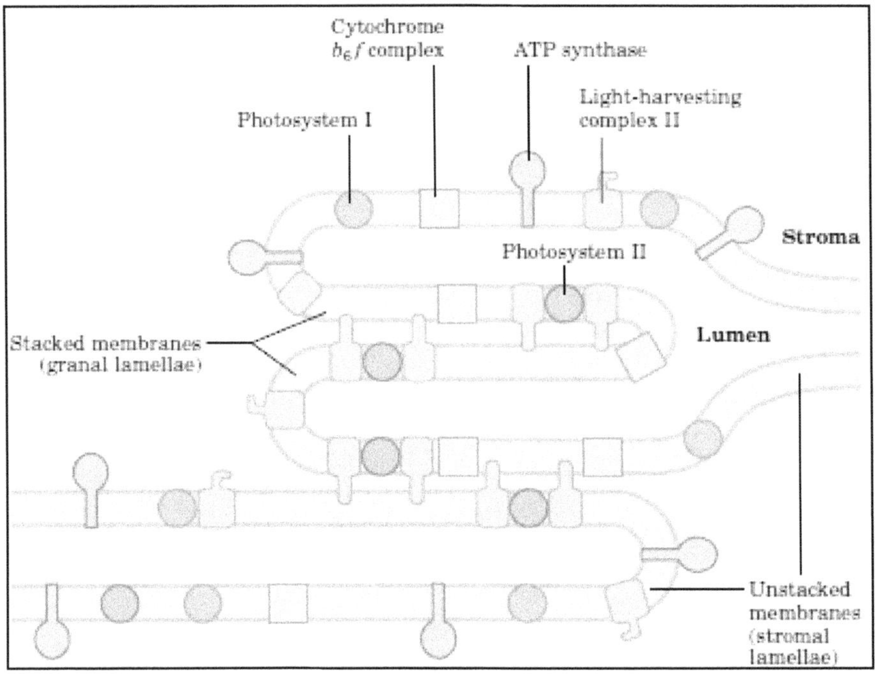

Figure 4.4: Lateral Heterogeneity in Thylakoid Membrane showing Distribution of PS I, PSII, Cyt b/f, and ATPase Complexes between Appressed (Stacked) and Non-appressed (Nonstacked) Region.

4.2. Photosynthetic Pigments

The energy of sunlight is first absorbed by the pigments of the plant. All pigments active in photosynthesis are found in the chloroplast. The chlorophylls and bacteriochlorophylls (pigments found in certain bacteria) are the typical pigments of photosynthetic organisms, but all organisms contain a mixture of more than one kind of pigment, each serving a specific function. There are 3 different categories of photosynthetic pigments are — (1) chlorophyll (2) Carotenoid (3) Phycobilins

Chlorophylls *a* and *b* are abundant in green plants, and *c* and *d* are found in some protists and cyanobacteria. A number of different types of bacteriochlorophyll have been found; type *a* is the most widely distributed. shows the distribution of pigments in different types of photosynthetic organisms. All chlorophylls have a porphyrin system in which 4 pyrrol (cyclic Tetrapyrrole) rings. The centre of tetrapyrrol is occupied by divalent magnesium. The loss of magnecium ion from chlorophyll results in the formation of a nongreen product Pheophytin (early electron accepter). This porphyrrin skeleton bears a 20 hydro carbon 'phytol chain'(derivative of the 5- carbon isoprene), which is bound in easter linkage to the pyrrol ring IV. The empirical formula of Chl.a is C H O N Mg.The different types of carotenoids found in photosynthetic organisms are all linear molecules with multiple conjugated double bonds (Figure 4.5B). Absorption bands in the 400 to 500 nm region give carotenoids their characteristic orange color. The color of carrots, for example, is due to the carotenoid ?-carotene, whose structure an

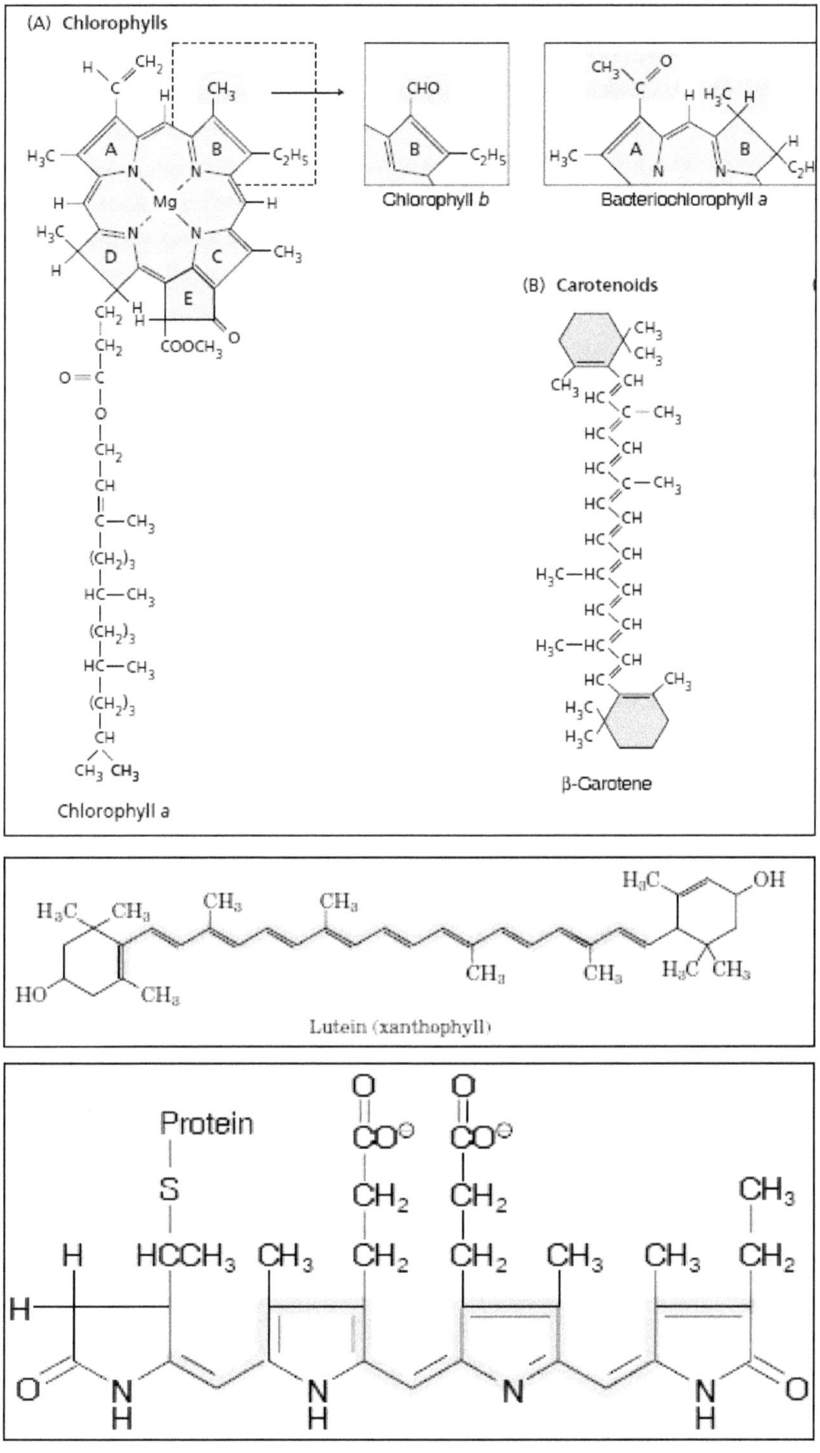

Figure 4.5: Structure of Several Photosynthetic Pigments.

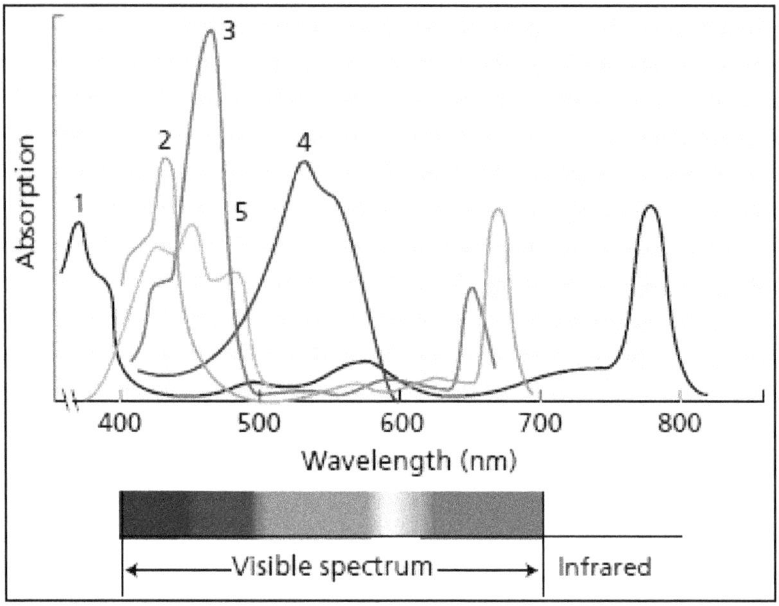

Figure 4.6: Absorption Spectra of some Photosynthetic Sigments.
1: Bacteriochlorophyll *a*; 2: Chlorophyll *a*; 3: Chlorophyll *b*; 4: Phycoerythrobilin; 5: Carotene (Courtesy of Taiz and Zieger).

absorption spectrum are shown in Figures 4.5 and 4.6 respectively. Carotenoids are found in all photosynthetic organisms. Carotenoids are integral constituents of the thylakoid membrane and are usually associated intimately with both antenna and reaction center pigment proteins. The light absorbed by the carotenoids is transferred to chlorophyll for photosynthesis; because of this role they are called accessory pigments. There are two major groups of carotenoids – the carotene and xanthophyll (such as lutein, zeaxanthin.and violoxanthin). The third category of photosynthetic pigments is Phycobilins. This is water soluble pigment found in blue-green algae and red algae. The phycobilins are three types – the phycoerythrin(red pigment) and phycocyanin(blue pigment) and Allophycocyanin. These pigments are open tetrapyrrol (chromophore) but donot contain Mg and phyol chain. It linked to special protein,forming phycobiliprotein (chromoprotein).These located in granular structure called phycobilisome.

4.3. Light

Light has properties of both particles and waves. A wave is characterized by a wavelength,denoted by the Greek letter lambda (*l*), which is the distance between successive wave crests. The frequency, represented by the Greek letter nu (*n*), is the number of wave crests that pass an observer in a given time. A simple equation relates the wavelength, the frequency, and the speed of any wave:

$$c = \lambda n$$

where c is the speed of the wave–in the present case, the speed of light (3.0 x?108 m s^{-1}). The light wave is a transverse(side-to-side) electromagnetic wave, in which both electric and magnetic fields oscillate perpendicularly to the direction of propagation of the wave and at 90° with respect to each other.

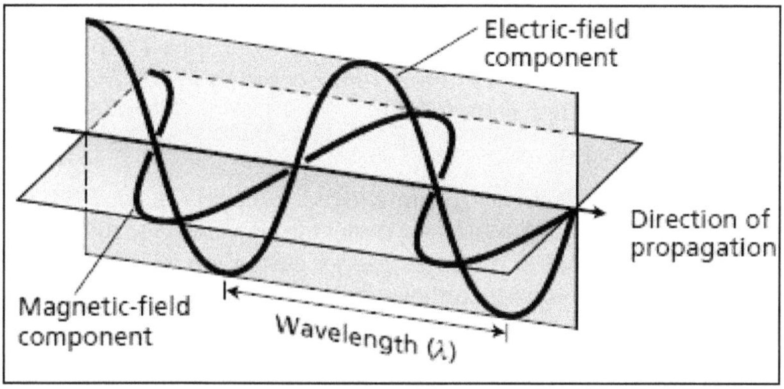

Figure 4.7: Light is a Transverse Electromagnetic Wave.

Light is also a particle, which we call a photon. Each photon contains an amount of energy that is called a quantum (plural *quanta*). The energy content of light is not continuous but rather is delivered in these discrete packets, the quanta. The energy (*E*) of a photon depends on the frequency of the light according to a relation known as Planck's law:

$$E = hc/\nu$$

where h is Planck's constant (6.626 ×~10′ J s). Sunlight is like a rain of photons of different frequencies.

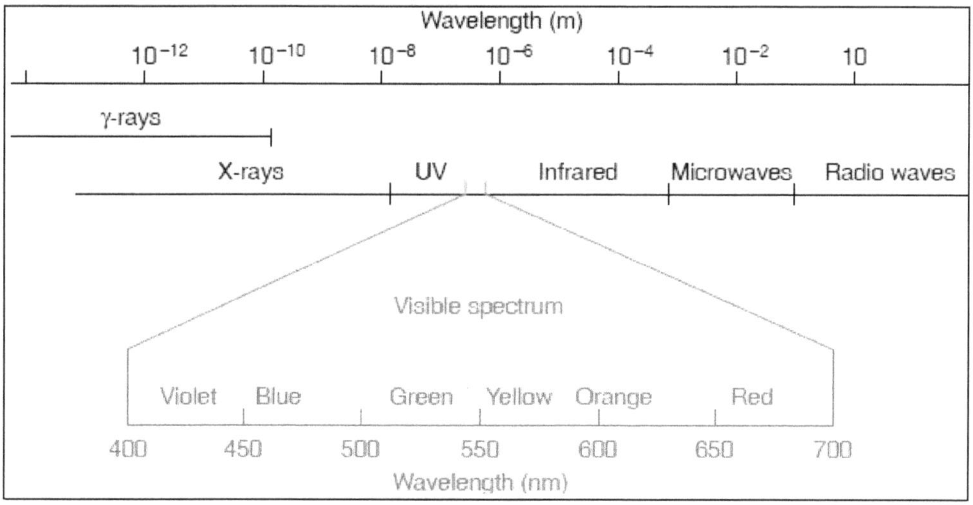

Figure 4.8: Electromagnetic Spectrum.
Wavelength (λ) and frequency (ν) are inversely related.

Our eyes are sensitive to only a small range of frequencies–the visible-light region of the electromagnetic spectrum.

Chlorophyll appears green to our eyes because it absorbs light mainly in the red and blue parts of the spectrum, so only some of the light enriched in green wavelengths (about 550 nm) is reflected into our eyes (see Figure 4.9). The absorption of light is represented by Equation 1.0, in which chlorophyll (Chl) in its lowest-energy, or ground, state absorbs a photon (represented by *h*) and makes a transition to a higher-energy, or excited, state (Chl*):

$$\text{Chl} + h \rightarrow \text{Chl}^* \quad (1.0)$$

The distribution of electrons in the excited molecule is somewhat different from the distribution in the groundstate molecule (Figure 4.9) Absorption of blue light excites the chlorophyll to a higher energy state than absorption of red light because the energy of photons is higher when their wavelength is shorter. In the higher excited state, chlorophyll is extremely unstable, very rapidly gives up some of its energy to the surroundings as heat, and enters the lowest excited state, where it can be stable for a maximum of several nanoseconds (10^{-9} s). Because of this inherent instability of the excited state, any process that captures its energy must be extremely rapid. In the lowest excited state, the excited chlorophyll has four alternative pathways for disposing of its available energy.

1. Excited chlorophyll can re-emit a photon and thereby return to its ground state–a process known as fluorescence. When it does so, the wavelength of fluorescence is slightly longer (and of lower energy) than the wavelength of absorption because a portion of the excitation energy is converted into heat before the fluorescent photon is emitted. Chlorophylls fluoresce in the red region of the spectrum.

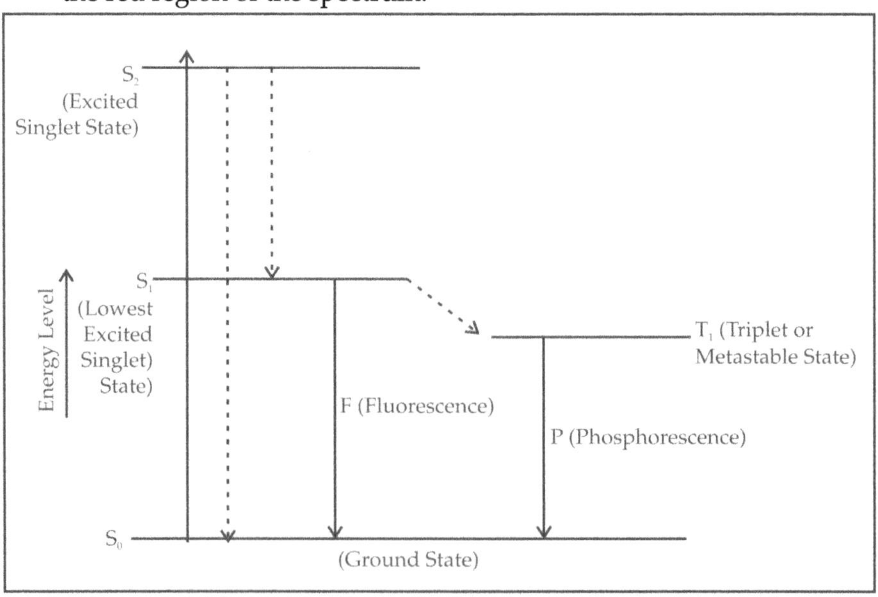

Figure 4.9: Diagram to Show Excitation of a Pigment Molecule to Fluorescence and Phosphorescence.

2. The excited chlorophyll can return to its ground state by directly converting its excitation energy into heat, with no emission of a photon.
3. Chlorophyll may participate in energy transfer, during which an excited chlorophyll transfers its energy to another molecule.
4. A fourth process is photochemistry, in which the energy of the excited state causes chemical reactions to occur. The photochemical reactions of photosynthesis are among the fastest known chemical reactions.

4.4. Action Spectra Relate Light Absorption to Photosynthetic Activity

The efficiency of light absorption and resulting physiological action can be displayed as a function of wavelength. An action spectrum depicts the magnitude of a response of a biological system to light, as a function of wavelength. For example, an action spectrum for photosynthesis can be constructed from measurements of oxygen evolution at different wavelengths (Figure 4.10). Often an action spectrum can identify the chromophore (pigment) responsible for a particular light-induced phenomenon. Absorption of different wavelengths of light, by a particular pigment, is plotted and called the absorption spectra of that pigment.Different type of pigments absorbs different wavelength of light.

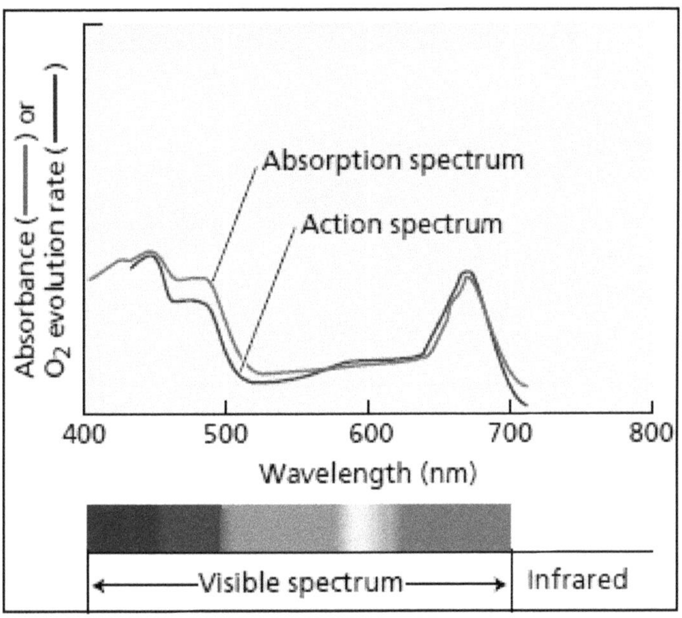

Figure 4.10: Absorption Spectra of Chlorophyll and Action Spectra of Photosynthesis (Courtesy of Taiz and Zieger).

Some of the first action spectra were measured by T. W. Engelmann in the late 1800s (Figure 4.11). Engelmann used a prism to disperse sunlight into a rainbow that was allowed to fall on an aquatic algal filament. A population of O_2-seeking bacteria was introduced into the system. The bacteria congregated in the regions of the filaments that evolved the most O_2. These were the regions illuminated by blue light and red light, which are strongly absorbed by chlorophyll. Today, action spectra can be measured in room-sized spectrographs in which a huge monochromator bathes the experimental samples in monochromatic light. But the principle of the experiment is the same as that of Engelmann's experiments. Action spectra were very important for the discovery of two distinct photosystems operating in O_2-evolving photosynthetic organisms.

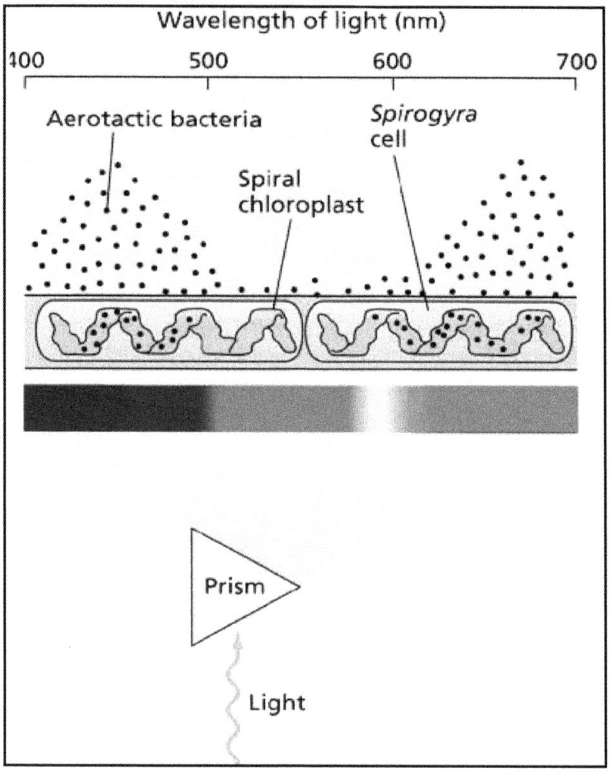

Figure 4.11: Diagram of the Action Spectrum Measurements by T. W. Engelmann. Engelmann projected a spectrum of light onto the spiral chloroplast of the filamentous green alga *Spirogyra*.

In 1932, Robert Emerson and William Arnold performed a key experiment that provided the first evidence for the cooperation of many chlorophyll molecules in energy conversion during photosynthesis. They delivered very brief (10–5 s) flashes of light to a suspension of the green alga *Chlorella pyrenoidosa* and measured the amount of oxygen produced. The flashes were spaced about 0.1 s apart, a time that Emerson and Arnold had determined in earlier work was long enough for the enzymatic steps of the process to be completed before the arrival of the next flash. The investigators varied the energy of the flashes and found that at high energies the oxygen production did not increase when a more intense flash was given: The photosynthetic system was saturated with light (Figure 4.12). In their measurement

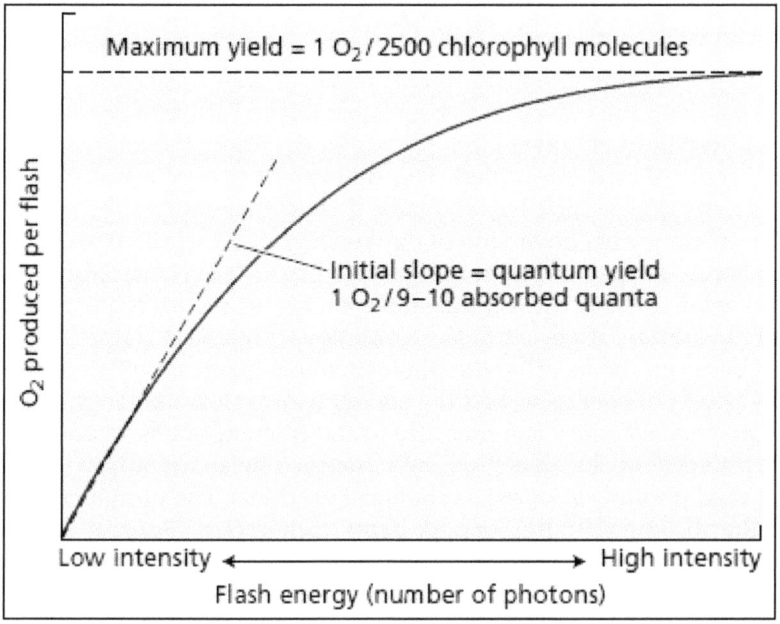

Figure 4.12: Relationship between Oxygen Production to Flash Energy.

of the relationship of oxygen production to flash energy, Emerson and Arnold were surprised to find that under saturating conditions, only one molecule of oxygen was produced for each 2500 chlorophyll molecules in the sample. We know now that several hundred pigments are associated with each reaction center and that each reaction center must operate four times to produce one molecule of oxygen–hence the value of 2500 chlorophylls per O_2. The reaction centers and most of the antenna complexes are integral components of the photosynthetic membrane. In eukaryotic photosynthetic organisms, these membranes are found within the chloroplast; in photosynthetic prokaryotes, the site of photosynthesis is the plasma membrane or membranes derived from it. The graph shown in Figure 4.12 permits us to calculate another important parameter of the light reactions of photosynthesis, the quantum yield. The quantum yield of photosynthesis (f) is defined as follows:

$$\phi = \frac{\text{Number of photochemical product}}{\text{Total number of quanta absorbed}}$$

In photosynthesis, for example, product yield would be measured as the amount of CO_2 taken up or O_2 evolved. In the linear portion (low light intensity) of the curve, an increase in the number of photons stimulates a proportional increase in oxygen evolution. Thus the slope of the curve measures the quantum yield for oxygen production. The quantum yield for a particular process can range from 0 (if that process does not respond to light) to 1.0 (if every photon absorbed contributes to the process). In functional chloroplasts kept in dim light, the quantum yield of photochemistry is approximately 0.95, the quantum yield of fluorescence is 0.05 or lower, and the quantum yields of other processes are negligible. The vast majority of excited chlorophyll molecules therefore lead to photochemistry.

Although the photochemical quantum yield under optimum conditions is nearly 100 per cent, the *efficiency* of the conversion of light into chemical energy is much less. If red light of wavelength 680 nm is absorbed, the total energy input is 1760 kJ per mole of oxygen formed. This amount of energy is more than enough to drive the reaction, which has a standard state free-energy change of +467 kJ mol–1. The efficiency of conversion of light energy at the optimal wavelength into chemical energy is therefore about 27 per cent, which is remarkably high for an energy conversion system. Most of this stored energy is used for cellular maintenance processes; the amount diverted to the formation of biomass is much less.There is no conflict between the fact that the photochemical quantum efficiency (quantum yield) is nearly 1 (100 per cent) and the energy conversion efficiency is only 27 per cent. The *quantum efficiency* is a measure of the fraction of absorbed photons that engage in photochemistry; the *energy efficiency* is a measure of how much energy in the absorbed photons is stored as chemical products. The numbers indicate that almost all the absorbed photons engage in photochemistry, but only about a fourth of the energy in each photon is stored, the remainder being converted to heat.

One of the experiments carried out by Emerson, measured the quantum yield of photosynthesis as a function of wavelength and revealed an effect known as the red drop (Figure 4.13). If the quantum yield is measured for the wavelengths at which chlorophyll absorbs light, the values found throughout most of the range are fairly constant, indicating that any photon absorbed by chlorophyll or other pigments is as effective as any other photon in driving photosynthesis. However, the yield drops dramatically in the far-red region of chlorophyll absorption (greater than 680 nm). This drop cannot be caused by a decrease in chlorophyll absorption because the quantum yield measures only light that has actually been absorbed. Thus, light with a wavelength greater than 680 nm is much less efficient than light of shorter wavelengths. Another puzzling experimental result was the enhancement

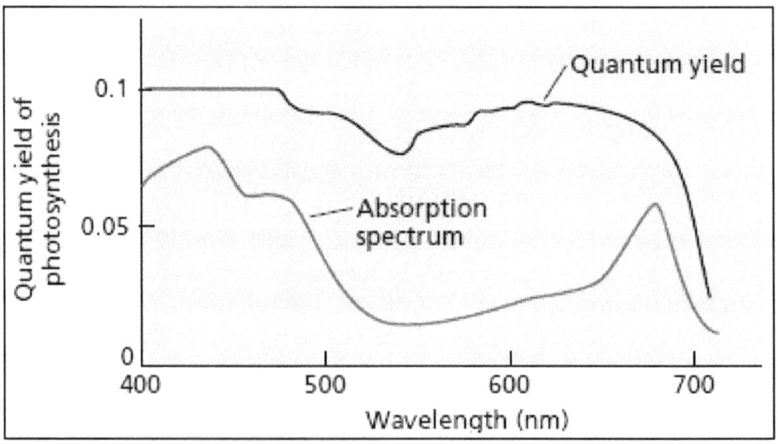

Figure 4.13: Red Drop Effect.

effect, also discovered by Emerson. He measured the rate of photosynthesis separately with light of two different wavelengths and then used the two beams simultaneously (Figure 4.14). When red and far-red light were given together, the rate of photosynthesis was greater than the sum of the individual rates.

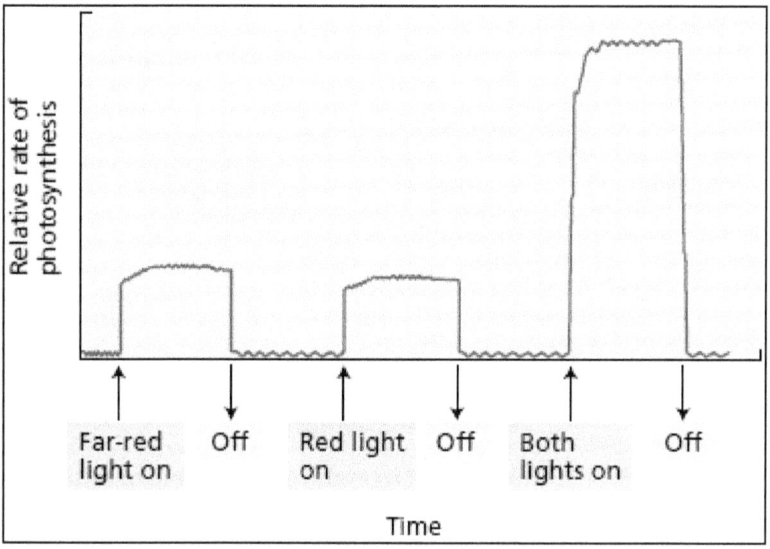

Figure 4.14: Enhancement Effect.

4.5. Photosystems Complex

Photosystem - <
- Antenna complex (*i.e.* LHC, chlorophylls other accessory pigments and associated proteins)
- Core complex (*i.e.* reaction centre, protein containing chlorophylls and electron donars and acceptors)

A portion of the light energy absorbed by chlorophylls and carotenoids is eventually stored as chemical energy via the formation of chemical bonds. This conversion of energy from one form to another is a complex process that depends on cooperation between many pigment molecules and a group of electron transfer proteins.The majority of the pigments serve as an antenna complex, collecting light and transferring the energy to the reaction center complex, where the chemical oxidation and reduction reactions leading to long-term energy storage take place (Figure 4.15).

Photosystems I and II (PSI and PSII), operate in series to carry out the early energy storage reactions of photosynthesis. Photosystem I preferentially absorbs far-red light of wavelengths greater than 680 nm; photosystem II preferentially absorbs red light of 680 nm and is driven very poorly by far-red light. This wavelength dependence explains the enhancement effect and the red drop effect. Another difference between the photosystems is that;

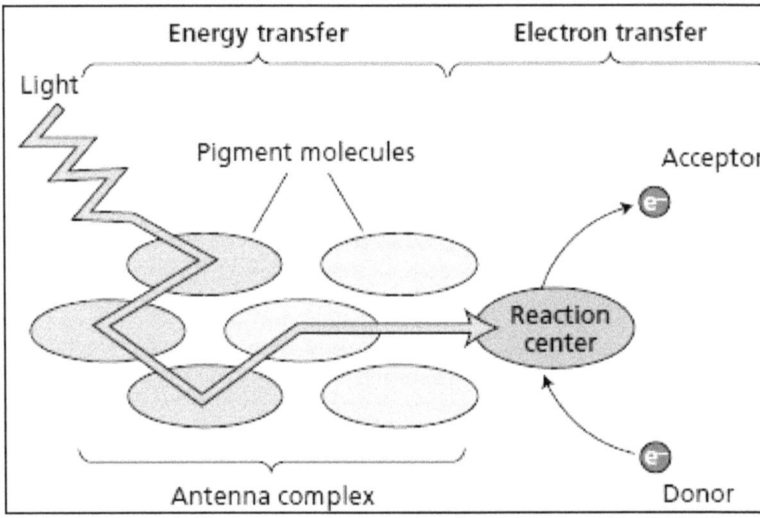

Figure 4.15: Diagrammatic Representation of Energy Transfer during Photosynthesis.

- Photosystem I produces a strong reductant, capable of reducing $NADP^+$, and a weak oxidant.
- Photosystem II produces a very strong oxidant, capable of oxidizing water, and a weaker reductant than the one produced by photosystem I.

The reductant produced by photosystem II re-reduces the oxidant produced by photosystem I. These properties of the two photosystems are shown schematically in Figure 1.17. The scheme of photosynthesis depicted in Figure 1.17, called the *Z* (for *zigzag*) *scheme*, has become the basis for understanding O_2-evolving (oxygenic) photosynthetic organisms. It accounts for the operation of two physically and chemically distinct photosystems (I and II), each with its own antenna pigments and photochemical reaction center. The two photosystems are linked by an electron transport chain

PSI and PSII have distinct absorption characteristics. The reaction center chlorophyll is transiently in an oxidized state after losing an electron and before being re-reduced by its electron donor. In the oxidized state, the strong light absorbance in the red region of the spectrum that is characteristic of chlorophylls is lost, or bleached. It is therefore possible to monitor the redox state of these chlorophylls by time-resolved optical absorbance measurements in which this bleaching is monitored directly. Using such techniques, Bessel Kok found that the reaction center chlorophyll of photosystem I absorbs maximally at 700 nm in its reduced state. Accordingly, this chlorophyll is named P700 (the P stands for *pigment*). H. T. Witt and coworkers found the analogous optical transient of photosystem II at 680 nm, so its reaction center chlorophyll is known as P680. The primary donor of photosystem I, P700, is a dimer of chlorophyll *a* molecules. Photosystem II also contains a dimer of chlorophylls, although the primary donor, P680, may not reside entirely on these pigments. In the oxidized state, reaction center chlorophylls contain an unpaired electron.

Photosystem II

Photosystem II is contained in a multisubunit protein supercomplex (Figure 4.17) (Barber *et al.*, 1999). In higher plants, the multisubunit protein supercomplex has two complete reaction centers and some antenna complexes. The core of the reaction center consists of two membrane proteins known as D1 and D2, as well as other proteins and the antenna consists of two chlorophyll-proteins (CP) known as CP 43 and CP 47 as shown in Figure 4.17 (Zouni *et al.*, 2001). The chlorophyll protein(CP) inserted in to membrane, in many cases Histidine residues in the protein serve as ligands to Mg in a chlorophyll. The primary donor chlorophyll (P680), additional chlorophylls, carotenoids, pheophytins, and plastoquinones are bound to the membrane proteins D1 and D2. These proteins have some sequence similarity to the L and M peptides of purple bacteria. Other proteins serve as antenna complexes or are involved in oxygen evolution. Some, such as cytochrome *b*559, have no known function but may be involved in a protective cycle around photosystem II. Protein components of PS II are given in Figure 4.16.

Protein	Molecular mass (kDa)	Localization	Encoded in	Function
D_1	32	In membrane	Chloroplast	Binding of P_{680}, Pheo, Q_B, Tyr, Mn-cluster
D_2	34	″	″	Binding of P_{680}, Pheo, Q_A, Mn-cluster
CP_{47}	47	″	″	Core-antenna, binds peripheral antennae LHC
CP_{43}	43	″	″	″ ″
Cyt-$b_{559\alpha}$	9	″	″	Binds heme, protection of PS II against light damage
Cyt-$b_{559\beta}$	4	″	″	″
Manganese-stabilizing protein (MSP)	33	Peripheral: lumen	Nucleus	Stabilization of Mn-cluster
P	23	″	″	?
Q	16	″	″	?
R	10	Peripheral: stroma	″	?

Figure 4.16: Photosystem II Multisubunit Protein Supercomplex.

The physical mechanism by which excitation energy is conveyed from the chlorophyll that absorbs the light to the reaction center is thought to be resonance transfer. By this mechanism the excitation energy is transferred from one molecule to another by a non radiative process.

The sequence of pigments within the antenna that funnel absorbed energy toward the reaction center has absorption maxima that are progressively shifted toward longer red wavelengths. This red shift in absorption maximum means that the energy of the excited state is somewhat lower nearer the reaction center than in the more peripheral portions of the antenna system. As a result of this arrangement, when excitation is transferred, for example, from a chlorophyll *b* molecule absorbing maximally at 650 nm to a chlorophyll *a* molecule absorbing maximally at 670 nm, the difference in energy between these two excited chlorophylls is lost to the environment as heat.

In all eukaryotic photosynthetic organisms that contain both chlorophyll *a* and chlorophyll *b*, the most abundant antenna proteins are members of a large family of structurally related proteins. Some of these proteins are associated primarily with photosystem II and are called light-harvesting complex II (LHCII) proteins; others are associated with photosystem I and are called *LHCI* proteins. These antenna complexes are also known as chlorophyll *a/b* antenna proteins (Paulsen 1995; Green and Durnford 1996. The protein contains three -helical regions and binds about 15 chlorophyll *a* and *b* molecules, as well as a few carotenoids. Only some of these pigments are visible in the resolved structure. The structure of the LHCI proteins has not yet been determined but is relatively small and has a chlorophyll a/b ratio of about 4/1 and appears tightly bound to core photosystem. while the LHCII proteins contain 50-60 per cent of total chlorophyll and chlorophyll a/b ratio of about ½. LHC II has no reaction centre and is not necessary for the survival of the plant. All of these proteins have significant sequence similarity and are almost certainly descendants of a common ancestral protein (Grossman *et al.*, 1995; Green and Durnford 1996). Light absorbed by carotenoids or chlorophyll *b* in the LHC proteins is rapidly transferred to chlorophyll *a* then to other antenna pigments that are intimately associated with the reaction center. The LHCII complex is also involved in regulatory processes.

4.6. Mechanism of Light Reaction (or Energy transducer or Hill reaction or Photophosphorylation)

Photophosphorylation

Light-driven production of ATP by chloroplast is known as photophosphorylation. Formation of ATP in association with noncyclic electron transport is known as noncyclic photophosphorylation while cyclic electron transport will also support ATP synthesis, a process known as cyclic photophosphorylation

"Z" Scheme or Noncyclic Photophosphorylation or Linear Electron Transport

In Z- scheme (Figure 4.17), in which all the electron carriers known to function in electron flow from H_2O to $NADP^+$ are arranged vertically at their midpoint redox potentials. Components known to react with each other are connected by arrows, so the Z scheme is really a synthesis of both kinetic and thermodynamic information. The large vertical arrows represent the input of light energy into the system. Photons excite the specialized chlorophyll of the reaction centers (P680 for PSII, and P700 for PSI), and an electron is ejected. The electron then passes through a series of electron carriers and eventually reduces P700 (for electrons from PSII) or NADP+ (for electrons from PSI). Much of the following discussion describes the journeys of these electrons and the nature of their carriers.

Almost all the chemical processes that make up the light reactions of photosynthesis are carried out by four major protein complexes: photosystem II, the cytochrome *b6 f* complex, photosystem I, and the ATP synthase. These four

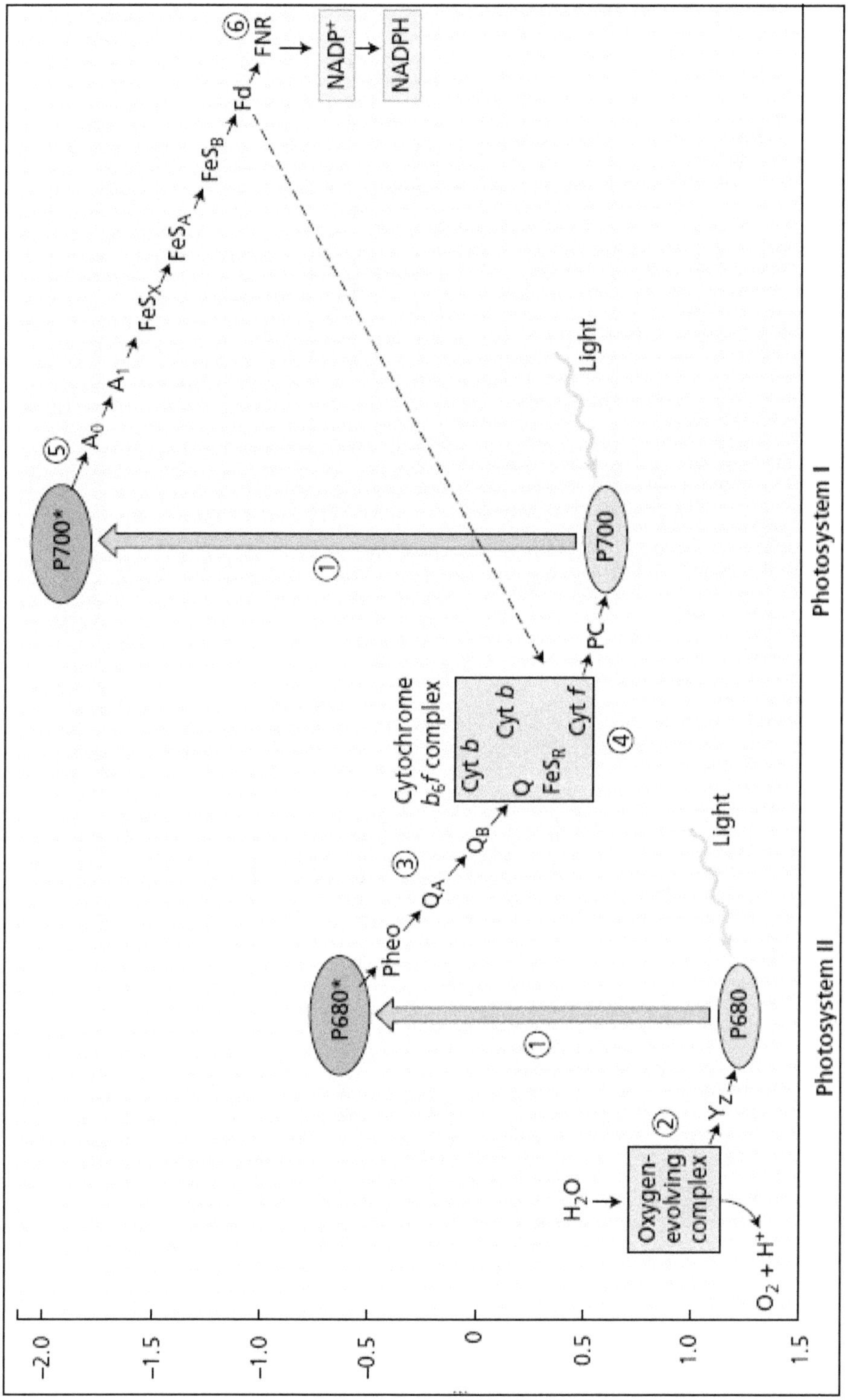

Figure 4.17: Diagram of Z-scheme (Courtesy of Taiz and Zieger).

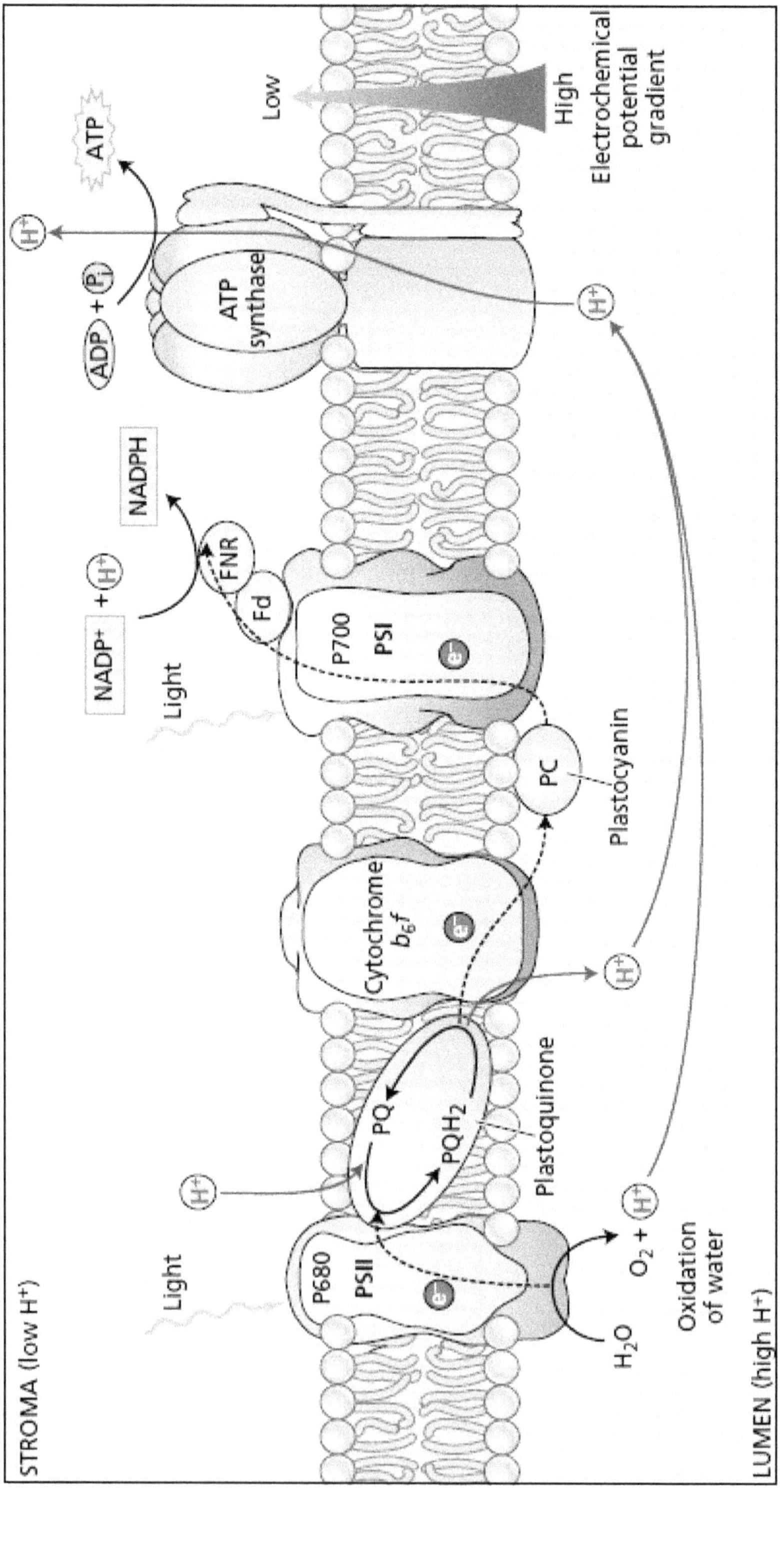

Figure 4.18: The Transfer of Electrons and Protons in the Thylakoid Membrane is carried out Vectorially by Four Protein Complex (Courtesy of Taiz and Zieger).

integral membrane complexes are vectorially oriented in the thylakoid membrane to function as follows (Figure 4.17):

- ✰ Photosystem II oxidizes water to O_2 in the thylakoid lumen and in the process releases protons into the lumen.
- ✰ Cytochrome *b6 f* receives electrons from PSII and delivers them to PSI. It also transports additional protons into the lumen from the stroma.
- ✰ Photosystem I reduces $NADP^+$ to NADPH in the stroma by the action of ferredoxin (Fd) and the flavoprotein ferredoxin–NADP reductase (FNR).
- ✰ ATP synthase produces ATP as protons diffuse back through it from the lumen into the stroma.

4.7. Photolysis of Water

Water is oxidized according to the following chemical reaction

$$2\,H_2O \rightarrow O_2 + 4\,H^+ + 4\,e^-$$

This equation indicates that four electrons are removed from two water molecules, generating an oxygen molecule and four hydrogen ions. (For more on oxidation–reduction reactionsWater is a very stable molecule. Oxidation of water to form molecular oxygen is very difficult, and the photosynthetic oxygen-evolving complex is the only known biochemical system that carries out this reaction. Photosynthetic oxygen evolution is also the source of almost all the oxygen in Earth's atmosphere. The chemical mechanism of photosynthetic water oxidation is not yet known, although many studies have provided a substantial amount of information about the process (Figure 4.18). The protons produced by water oxidation are released into the lumen of the thylakoid, not directly into the stromal compartment (see Figure 4.19). They are released into the lumen because of the vectorial nature of the membrane and the fact that the oxygen-evolving complex is localized on the interior surface of the thylakoid. These protons are eventually transferred from the lumen to the stroma by translocation through ATP synthase. In this way the protons released during water oxidation contribute to the electrochemical potential driving ATP formation. It has been known for many years that manganese (Mn) is an essential cofactor in the water-oxidizing process, and a classic hypothesis in photosynthesis research postulates that Mn ions undergo a series of oxidations–which are known as *S states,* and are labeled S0, S1, S2, S3, and S4 –that are perhaps linked to H_2O oxidation and the generation of O_2 (see Figure 4.19). Analytical experiments indicate that four Mn ions are associated with each oxygen-evolving complex. Other experiments have shown that Cl^- and Ca^{2+} ions are essential for O_2 evolution (see Figure 4.19). One electron carrier, generally identified as Yz, functions between the oxygen-evolving complex and P680 (see Figures 4.19). To function in this region, Yz needs to have a very strong tendency to retain its electrons. This species has been identified as a radical formed from a tyrosine residue in the D1 protein of the PSII reaction center.

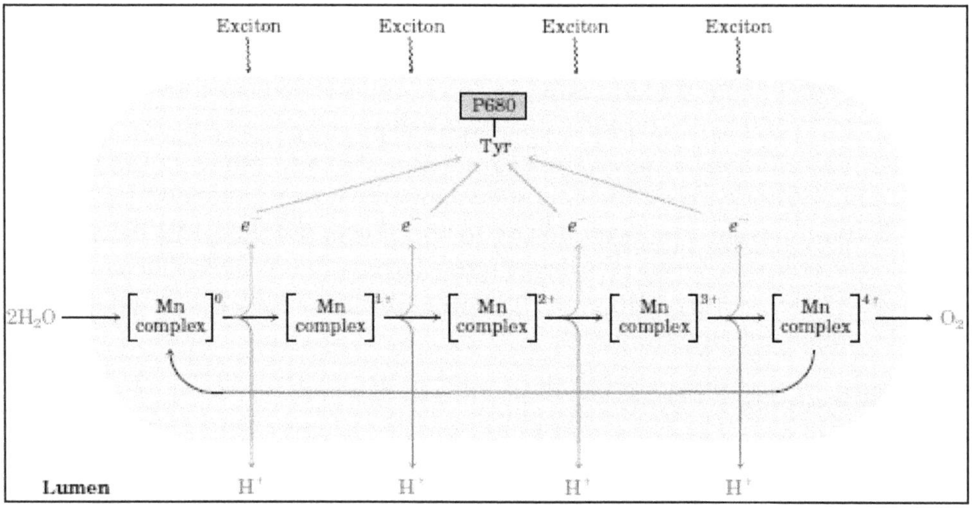

Figure 4.19: Water Splitting Activity of the Oxygen Evolving Complex.

Pheophytin and Two Quinones Accept Electrons from Photosystem II

Pheophytin acts as an early acceptor in photosystem II, followed by a complex of two plastoquinones in close proximity to an iron atom. Pheophytin is chlorophyll in which the central magnesium atom has been replaced by two hydrogen atoms. This chemical change gives pheophytin chemical and spectral properties that are slightly different from those of chlorophyll. Two plastoquinones (QA and QB) are bound to the reaction center and receive electrons from pheophytin in a sequential fashion (Okamura *et al.,* 2000). Transfer of the two electrons to QB reduces it to QB_2–, and the reduced QB2– takes two protons from the stroma side of the medium, yielding a fully reduced plastohydroquinone (QH_2) (Figure 4.20). The plastohydroquinone then dissociates from the reaction center complex and enters the hydrocarbon portion of the membrane, where it in turn transfers its electrons to the cytochrome *b6 f* complex. Unlike the large protein complexes of the thylakoid membrane, hydroquinone is a small, nonpolar molecule that diffuses readily in the nonpolar core of the membrane bilayer.

4.8. Electron Flow through the Cytochrome b6f Complex also Transports Protons

The cytochrome *b6 f* complex is a large multisubunit protein with several prosthetic groups (Cramer *et al.,* 1996; Berry *et al.,* 2000). It contains two *b*-type hemes and one *c*type heme (cytochrome *f*). In *c*-type cytochromes the heme is covalently attached to the peptide; in *b*-type cytochromes the chemically similar protoheme group is not covalently attached.In addition, the complex contains a Rieske iron–sulfur protein (named for the scientist who discovered it), in which two iron atoms are bridged by two sulfur atoms. The structures of cytochrome *f* and the related cytochrome *bc*1 complex have been determined and suggest a mechanism for electron and proton flow. The precise way by which electrons and protons flow through the cytochrome *b6 f* complex is not yet fully understood, but a mechanism

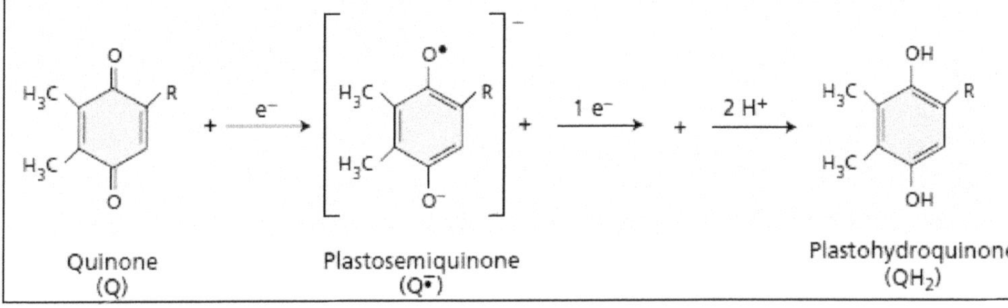

Figure 4.20: Reactions of Plastoquinone that Operate in Photosystem II.

known as the Q cycle accounts for most of the observations. In this mechanism, plastohydroquinone (QH_2) is oxidized, and one of the two electrons is passed along a linear electron transport chain toward photosystem I, while the other electron goes through a cyclic process that increases the number of protons pumped across the membrane (Figure 4.21). In the linear electron transport chain, the oxidized Rieske protein (FeSR) accepts an electron from plastohydroquinone (QH_2) and transfers it to cytochrome *f* (see Figure 4.21A). Cytochrome *f* then transfers an electron to the blue-colored copper protein plastocyanin (PC), which in turn reduces oxidized P700 of PSI. In the cyclic part of the process (see Figure 4.21), the plastosemiquinone. transfers its other electron to one of the *b*-type hemes, releasing both of its protons to the lumenal side of the membrane. The *b*-type heme transfers its electron through the second *b*-type heme to an oxidized quinone molecule, reducing it to the semiquinone form near the stromal surface ofthe complex. Another similar sequence of electron flow fully reduces the plastoquinone, which picks up protons from the stromal side of the membrane and is released from the *b6 f* complex as plastohydroquinone. The net result of two turnovers of the complex is that two electrons are transferred to P700, two plastohydroquinones are oxidized to the quinone form, and one oxidized plastoquinone is reduced to the hydroquinone form.

In addition, four protons are transferred from the stromal to the lumenal side of the membrane. By this mechanism, electron flow connecting the acceptor side of the PSII reaction center to the donor side of the PSI reaction center also gives rise to an electrochemical potential across the membrane, due in part to H^+ concentration differences on the two sides of the membrane. This electrochemical potential is used to power the synthesis of ATP. The cyclic electron flow through the cytochrome *b* and plastoquinone increases the number of protons pumped per electron beyond what could be achieved in a strictly linear sequence.

4.9. Plastoquinone and Plastocyanin Carry Electrons between Photosystem I and II

The location of the two photosystems at different sites on the thylakoid membranes (see Figure 4.4) requires that at least one component be capable of moving along or within the membrane in order to deliver electrons produced by photosystem II to photosystem I. The cytochrome *b6 f* complex is distributed

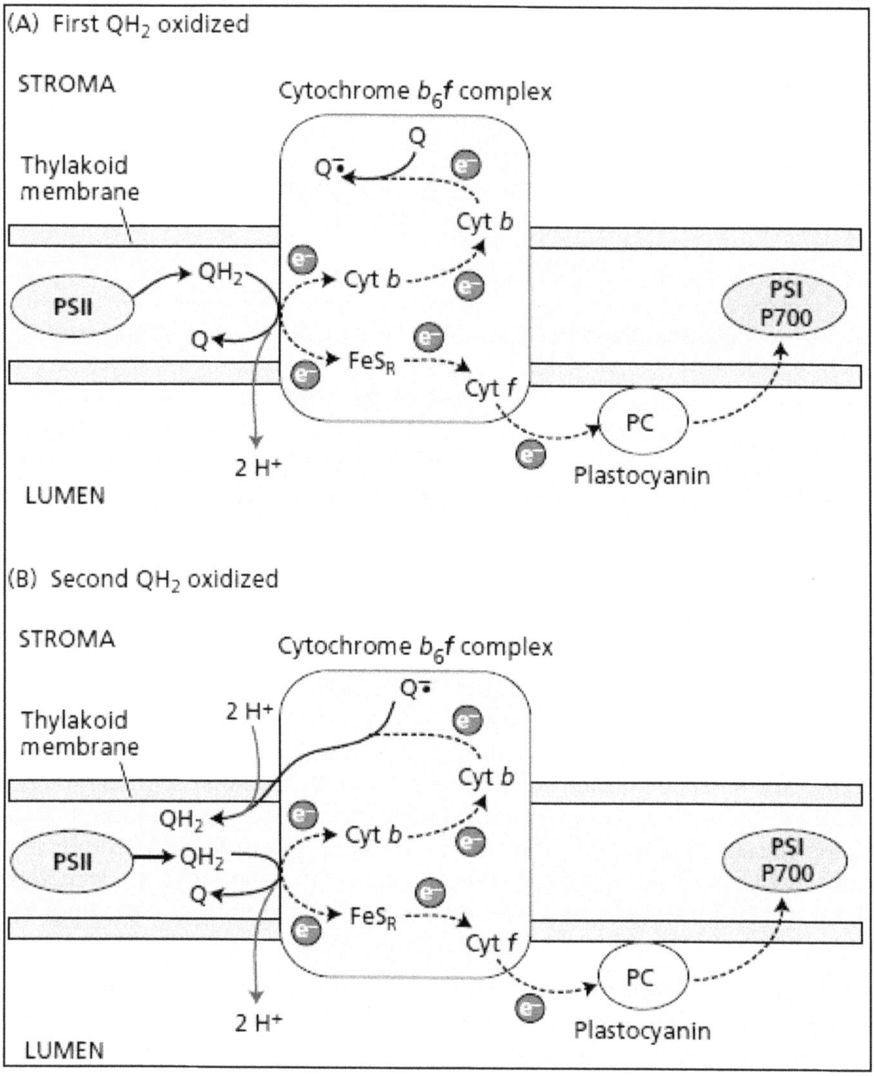

Figure 4.21: Mechanism of Electron and Proton Transfer in the Cytochrome b f Complex.

equally between the grana and the stroma regions of the membranes, but its large size makes it unlikely that it is the mobile carrier. Instead, plastoquinone or plastocyanin or possibly both are thought to serve as mobile carriers to connect the two photosystems. Plastocyanin is a small (10.5 kDa), water-soluble, copper-containing protein that transfers electrons between the cytochrome *b6 f* complex and P700. This protein is found in the lumenal space (see Figure 4.19). In certain green algae and cyanobacteria, a *c*-type cytochrome is sometimes found instead of plastocyanin; which of these two proteins is synthesized depends on the amount of copper available to the organism.

4.10. The Photosystem I Reaction Center Reduces $NADP^+$

The PSI reaction center complex is a large multisubunit complex (Figure 4.23) (Jordan *et al.*, 2001). In contrast to PSII, a core antenna consisting of about 100 chlorophylls is a part of the PSI reaction center, P700. The core antenna and P700 are bound to two proteins, PsaA and PsaB, with molecular masses in the range of 66 to 70 kDa. The antenna pigments form a bowl surrounding the electron transfer cofactors, which are in the center of the complex. In their reduced form, the electron carriers that function in the acceptor region of photosystem I are all extremely strong reducing agents. These reduced species are very unstable and thus difficult to identify. Evidence indicates that one of these early acceptors is a chlorophyll molecule, and another is a quinone species, phylloquinone, also known as vitamin K1. Additional electron acceptors include a series of three membrane-associated iron–sulfur proteins, or bound ferredoxins, also known as Fe–S centers FeSX, FeSA, and FeSB (Figure 4.22). Fe–S center X is part of the P700-binding protein; centers

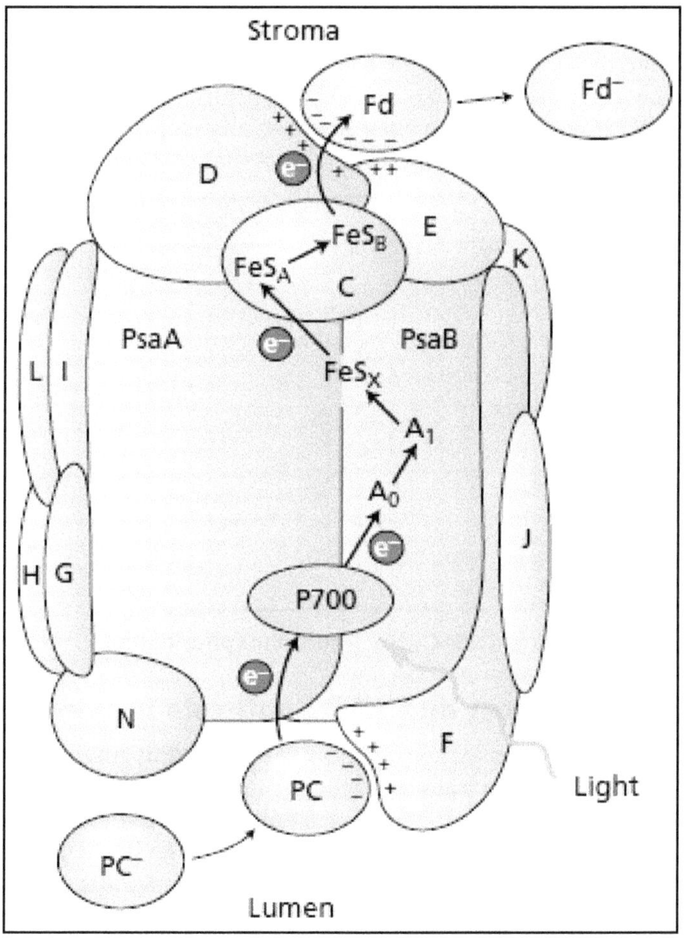

Figure 4.22: Structure of Photosystem I.

Aand B reside on an 8 kDa protein that is part of the PSI reaction center complex. Electrons are transferred through centers Aand B to ferredoxin (Fd), a small, water-soluble iron–sulfur protein (Figures 4.17 and 4.22). The membrane-associated flavoprotein ferredoxin–NADP reductase (FNR) reduces $NADP^+$ to NADPH, thus completing the sequence of noncyclic electron transport that begins with the oxidation of water (Karplus *et al.*, 1991). In addition to the reduction of $NADP^+$, reduced ferredoxin produced by photosystem I has several other functions in the chloroplast, such as the supply of reductants to reduce nitrate and the regulation of some of the carbon fixation enzymes.

The noncyclic electron transport is most important in photosynthesis at it supplies assimilatory power, in the form of NADPH and ATP for CO_2 assimilation and purifies the atmosphere air.

4.11. Cyclic Photophosphorylation or Cyclic Electron Transport

Some of the cytochrome *b6 f* complexes are found in the stroma region of the membrane, where photosystem I is located. Under certain conditions cyclic electron flow from the reducing side of photosystem I, through the *b6 f* complex and back to P700, is known to occur. This cyclic electron flow is coupled to proton pumping into the lumen, which can be utilized for ATP synthesis but does not oxidize water or reduce $NADP^+$. Cyclic electron flow is especially important as an ATP source in the bundle sheath chloroplasts of some plants that carry out C4 carbon fixation.

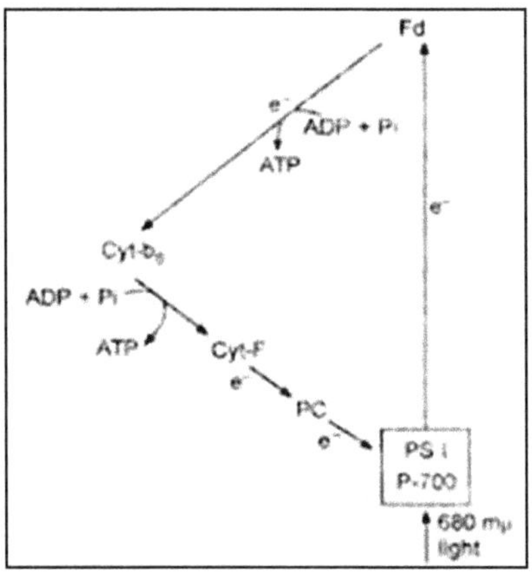

Figure 4.23: Cyclic Photophosphorylation

4.12. Proton Transport and ATP Synthesis in the Chloroplast

In the preceding sections we learned how captured light energy is used to reduce NADP+ to NADPH. Another fraction of the captured light energy is used for light-dependent ATP synthesis, which is known as photophosphorylation. This process was discovered by Daniel Arnon and his coworkers in the 1950s. In normal cellular conditions, photophosphorylation requires electron flow, although under some conditions electron flow and photophosphorylation can take place independently of each other. Electron flow without accompanying phosphorylation is said to be uncoupled. It is now widely accepted that photophosphorylation works via the chemiosmotic mechanism, first proposed in the 1960s by Peter Mitchell. Here we are concerned with chemiosmosis and transmembrane proton concentration differences used to make ATP in the chloroplast. The basic principle of chemiosmosis is that ion concentration differences and electric-potential differences across membranes are a source of free energy that can be utilized by the cell. As described by the second law of thermodynamics any nonuniform distribution of matter or energy represents a source of energy. Differences in chemical potential of any molecular species whose concentrations are not the same on opposite sides of a membrane provide such a source of energy. The asymmetric nature of the photosynthetic membrane and the fact that proton flow from one side of the membrane to the other accompanies electron flow. The direction of proton translocation is such that the stroma becomes more alkaline (fewer H^+ ions) and the lumen becomes more acidic (more H^+ ions) as a result of electron transport. Some of the early evidence supporting a chemiosmotic mechanism of photosynthetic ATP formation was provided by an elegant experiment carried out by André Jagendorf and coworkers

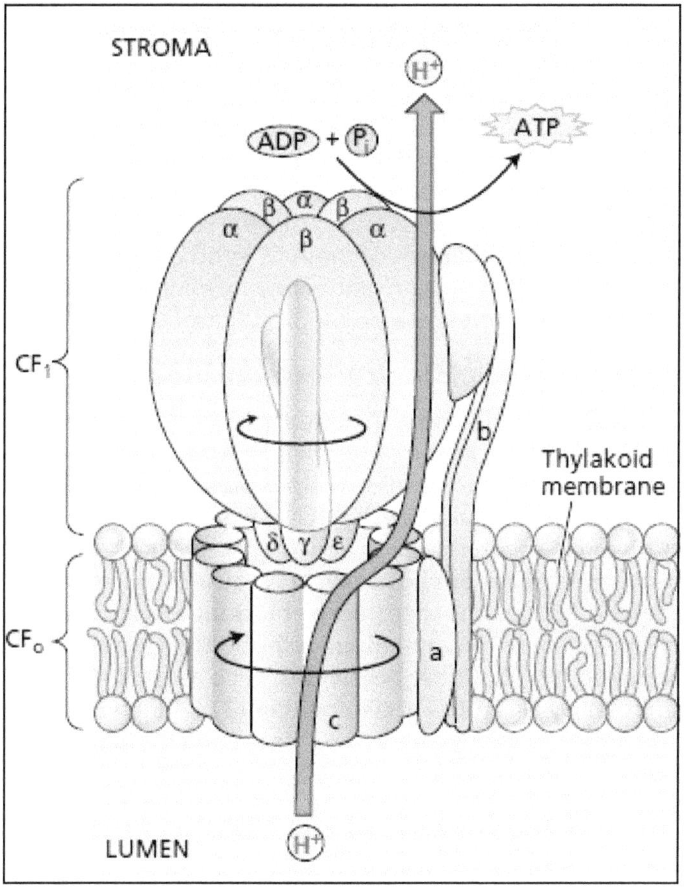

Figure 4.24: Structure of ATP Synthase, Consists of Multisubunits Complex CF_0–CF_1 (Courtesy of Taiz and Zieger).

(Figure 4.24). They suspended chloroplast thylakoids in a pH 4 buffer, and the buffer diffused across the membrane, causing the interior, as well as the exterior, of the thylakoid to equilibrate at this acidic pH. They then rapidly transferred the thylakoids to a pH 8 buffer, thereby creating a pH difference of 4 units across the thylakoid membrane, with the inside acidic relative to the outside. They found that large amounts of ATP were formed from ADP and Pi by this process, with no light input or electron transport. Mitchell proposed that the total energy available for ATP synthesis, which he called the proton motive force (*p*), is the sum of a proton chemical potential and a transmembrane electric potential. The stoichiometry of protons translocated per ATP synthesized has recently been found to be four H+ ions per ATP (Haraux and De Kouchkovsky 1998). In addition to the need for mobile electron carriers, the uneven distribution of photosystems II and I, and of ATP synthase at the thylakoid membrane (Figure 1.24), poses some challenges for the formation of ATP. ATP synthase is found only in the stroma lamellae and at the edges of the grana stacks. Protons pumped across the membrane by the cytochrome *b6 f* complex or protons produced by water oxidation in the middle of the grana must move laterally up to several tens of nanometers to reach ATP synthase. The ATP is synthesized by a large (400 kDa) enzyme complex known by several names: ATP synthase, ATPase (after the reverse reaction of ATP hydrolysis), and CFo–CF1 (Boyer 1997). This enzyme consists of two parts: a hydrophobic membrane-bound portion called CFo and a portion that sticks out into the stroma called CF1 (Figure 4.24). CFo appears to form a channel across the membrane through which protons can pass. CF1 is made up of several peptides; including three copies of each of the < and ®peptides arranged alternately much like the sections of an orange. Another remarkable aspect of the mechanism of the ATP synthase is that the internal stalk and probably much of the CFo portion of the enzyme rotate during catalysis (Yasuda *et al.,* 2001). The enzyme is actually a tiny molecular motor.

4.13. Some Herbicides Block Electron Flow

The use of herbicides to kill unwanted plants is widespread in moder n agriculture. Many different classes of herbicides have been developed, and they act by blocking amino acid, carotenoid, or lipid biosynthesis or by disrupting cell division. Other herbicides, such as DCMU (dichlorophenyldimethylurea) and paraquat, block photosynthetic electron flow (Figure 4.25). DCMU is also known as diuron. Paraquat has acquired public notoriety because of its use on marijuana crops. Many herbicides, DCMU among them, act by blocking electron flow at the quinone acceptors of photosystem II, by competing for the binding site of plastoquinone that is normally occupied by QB. Other herbicides, such as paraquat, act by accepting electrons from the early acceptors of photosystem I and then reacting with oxygen to form superoxide, O_2^-, a species that is very damaging to chloroplast components, especially lipids.

4.14. Repair and Regulation of the Photosynthetic Machinery

Photosynthetic systems face a special challenge. They are designed to absorb large amounts of light energy and process it into chemical energy. At the molecular level, the energy in a photon can be damaging, particularly under unfavorable

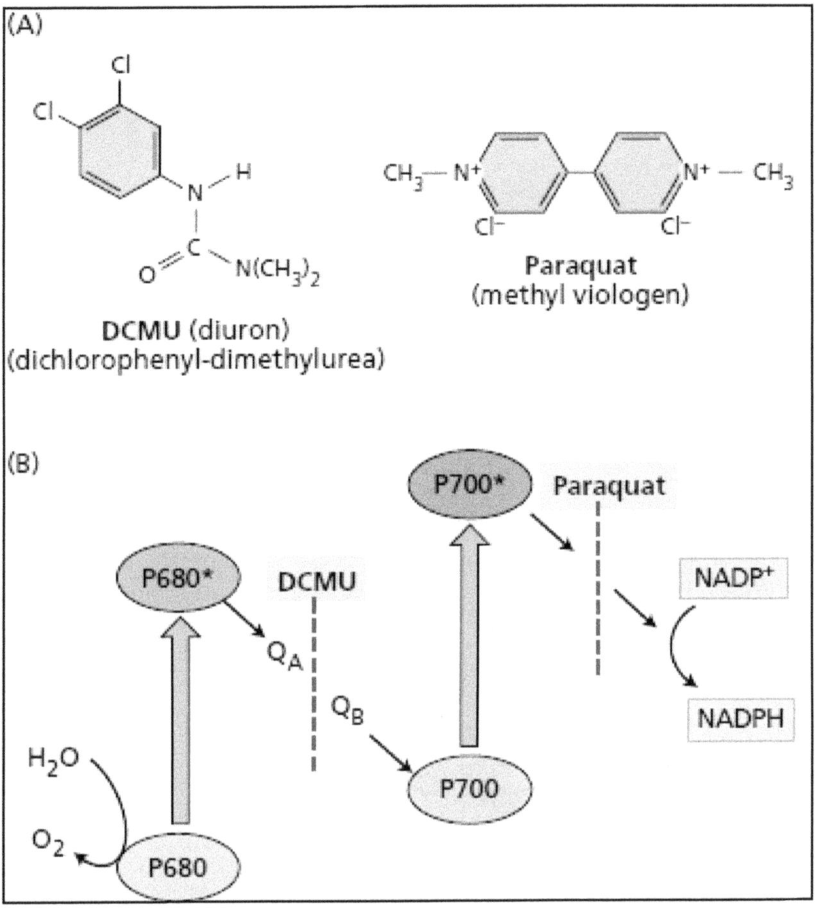

Figure 4.25: Chemical Structure and Mechanism of Action of Herbicide.

conditions. In excess, light energy can lead to the production of toxic species, such as superoxide, singlet oxygen, and peroxide, and damage can occur if the light energy is not dissipated safely (Horton *et al.*, 1996; Asada 1999; Müller *et al.*, 2001). Photosynthetic organisms therefore contain complex regulatory and repair mechanisms.

4.15. Carotenoids Serve as Photoprotective Agents

In addition to their role as accessory pigments, carotenoids play an essential role in photoprotection. The photosynthetic membrane can easily be damaged by the large amounts of energy absorbed by the pigments if this energy cannot be stored by photochemical reaction; this is why a protection mechanism is needed. The photoprotection mechanism can be thought of as a safety valve, venting excess energy before it can damage the organism. When the energy stored in chlorophylls in the excited state is rapidly dissipated by excitation transfer or photochemistry, the excited state is said to be quenched. If the excited state of chlorophyll is not

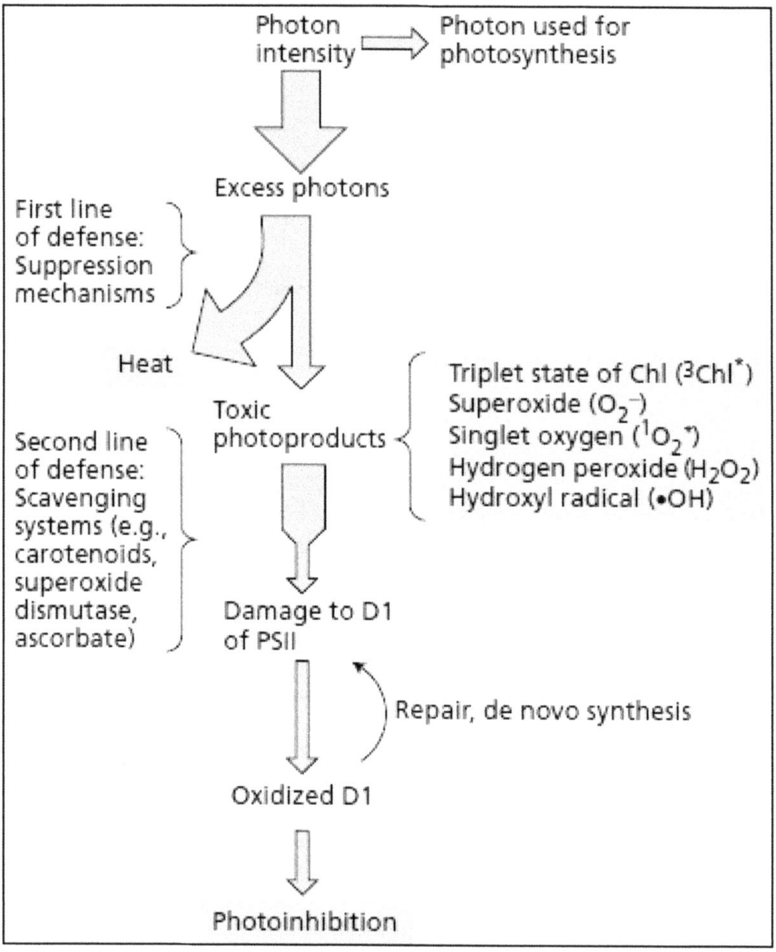

Figure 4.26: Overall Picture of Regulation of Photon Capture and the Protection and Repair of Photodamage.

rapidly quenched by excitation transfer or photochemistry, it can react with molecular oxygen to form an excited state of oxygen known as singlet oxygen ($1O_2$*). The extremely reactive singlet oxygen goes on to react with and damage many cellular components, especially lipids. Carotenoids exert their photoprotective action by rapidly quenching the excited state of chlorophyll. The excited state of carotenoids does not have sufficient energy to form singlet oxygen, so it decays back to its ground state while losing its energy as heat. Mutant organisms that lack carotenoids cannot live in the presence of both light and molecular oxygen–a rather difficult situation for an O_2-evolving photosynthetic organism. For non-O_2-evolving photosynthetic bacteria, mutants that lack carotenoids can be maintained under laboratory conditions if oxygen is excluded from the growth medium. Recently carotenoids were found to play a role in nonphotochemical quenching, which is a second protective and regulatory mechanism.

4.16. Some Xanthophylls also Participate in Energy Dissipation

Nonphotochemical quenching, a major process regulating the delivery of excitation energy to the reaction center, can be thought of as a "volume knob" that adjusts the flow of excitations to the PSII reaction center to a manageable level, depending on the light intensity and other conditions. The process appears to be an essential part of the regulation of antenna systems in most algae and plants. Nonphotochemical quenching is the quenching of chlorophyll fluorescence by processes other than photochemistry. As a result of nonphotochemical quenching, a large fraction of the excitations in the antenna system caused by intense illumination are quenched by conversion into heat (Krause and Weis, 1991). Nonphotochemical quenching is thought to be involved in protecting the photosynthetic machinery against overexcitation and subsequent damage. The molecular mechanism of nonphotochemical quenching is not well understood, although it is clear that the pH of the thylakoid lumen and the state of aggregation of the antenna complexes are important factors. Three carotenoids, called xanthophylls, are involved in nonphotochemical quenching: violaxanthin, antheraxanthin, and zeaxanthin (Figure

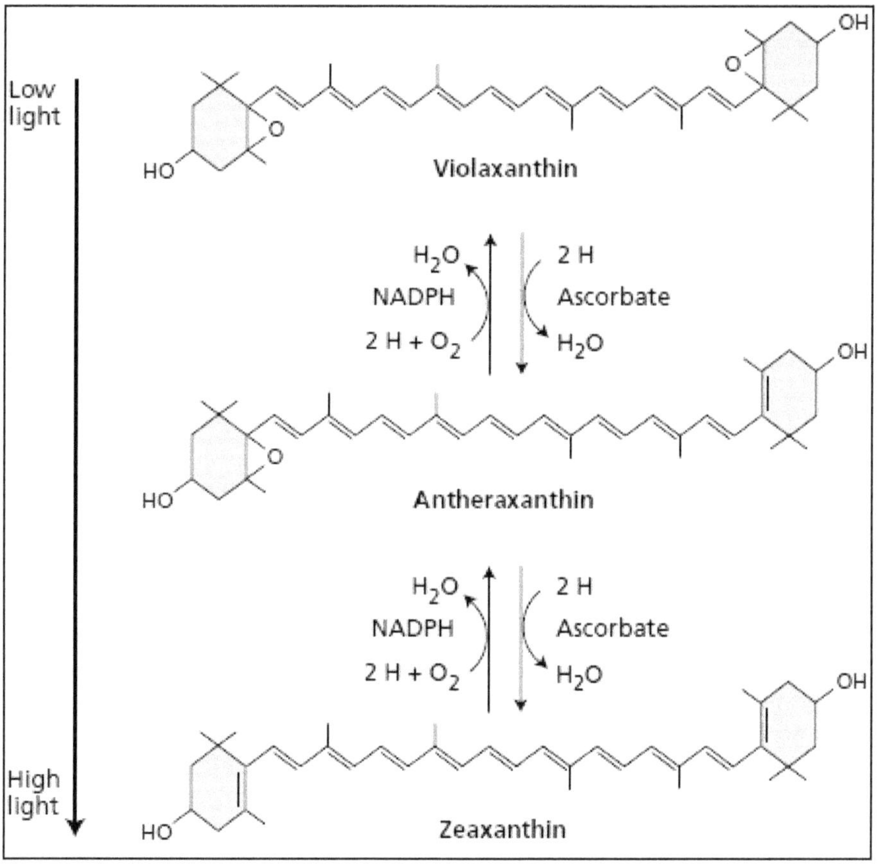

Figure 4.27: Chemical Structure of Violaxanthin, Antheraxanthin, and Zeaxanthin involved in Non-photochemical Quenching.

4.27). In high light, violaxanthin is converted into zeaxanthin, via the intermediate antheraxanthin, by the enzyme violaxanthin de-epoxidase. When light intensity decreases, the process is reversed. Binding of protons and zeaxanthin to light-harvesting antenna proteins is thought to cause conformational changes that lead to quenching and heat dissipation (Demmig Adams and Adams 1992; Horton *et al.*, 1996). Nonphotochemical quenching appears to be preferentially associated with a peripheral antenna complex of photosystem II, the PsbS protein (Li *et al.*, 2000).

4.17. Carbon Assimilation (Dark Reaction)

These stroma reactions were long thought to be independent of light and, as a consequence, were referred to as the *dark reactions*. However, because these stroma-localized reactions depend on the products of the photochemical processes, and are also directly regulated by light, they are more properly referred to as the *carbon reactions of photosynthesis*. The photochemical oxidation of water to molecular oxygen is coupled to the generation of ATP and reduced pyridine nucleotide (NADPH) by reactions taking place in the chloroplast thylakoid membrane. The reactions catalyzing the reduction of CO_2 to carbohydrate are coupled to the consumption of NADPH and ATP by enzymes found in the stroma.

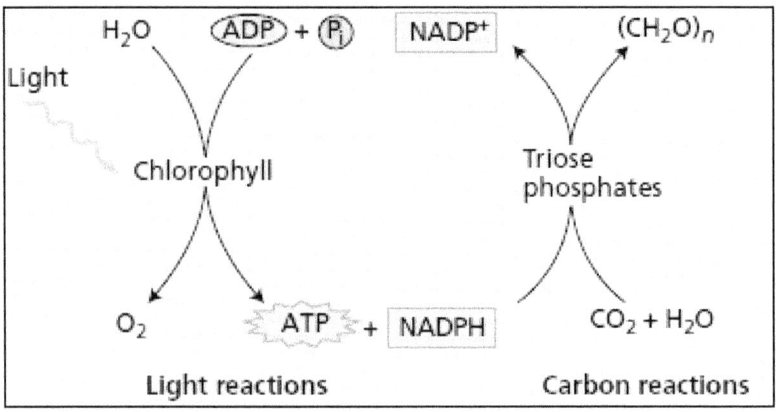

Figure 4.28: The Light and Carbon Reaction of Photosynthesis.

4.18. Calvin Cycle, or Reductive Pentose Phosphate [RPP] Cycle (C_3 Cycle)

In the Calvin cycle, CO_2 and water from the environment are enzymatically combined with a five-carbon acceptor molecule to generate two molecules of a three-carbon intermediate. This intermediate (3 phosphoglycerate) is reduced to carbohydrate by use of the ATP and NADPH generated photochemically. The cycle is completed by regeneration of the five-carbon acceptor (ribulose-1,5-bisphosphate, abbreviated RuBP).

The Calvin cycle proceeds in three stages (Figure 4.28):

1. *Carboxylation* of the CO_2 acceptor ribulose-1,5-bisphosphate, forming two molecules of 3-phosphoglycerate, the first stable intermediate of the Calvin cycle
2. *Reduction* of 3-phosphoglycerate, forming gyceraldehyde- 3-phosphate, a carbohydrate
3. *Regeneration* of the CO_2 acceptor ribulose-1,5-bisphosphate from glyceraldehyde-3-phosphate.

The carbon in CO_2 is the most oxidized form found in nature (+4). The carbon of the first stable intermediate, 3- phosphoglycerate, is more reduced (+3), and it is further reduced in the glyceraldehyde-3-phosphate product (+1). Overall, the early reactions of the Calvin cycle complete the reduction of atmospheric carbon and, in so doing, facilitate its incorporation into organic compounds.

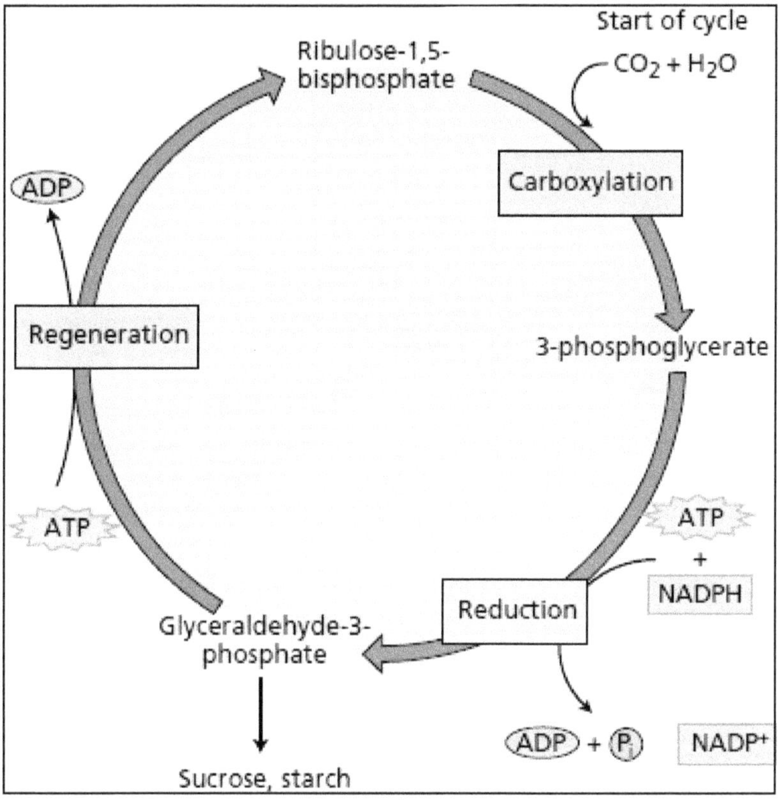

Figure 4.29a: Diagrammatic Representation of Calvin Cycle.

Carboxylation

CO_2 enters the Calvin cycle by reacting with ribulose-1,5- bisphosphate to yield two molecules of 3 phosphoglycerate (Figure 4.28), a reaction catalyzed by the chloroplast enzyme ribulose bisphosphate carboxylase/oxygenase, referred to as rubisco. The enzyme also has an oxygenase activity in which O_2 competes with CO_2 for the common substrate ribulose 1,5-bisphosphate (Lorimer 1983). This property limits net CO_2 fixation. CO_2 is added to carbon 2 of ribulose- 1,5-bisphosphate, yielding an unstable, enzyme-bound intermediate, which is hydrolyzed to yield two molecules of the stable product 3-phosphoglycerate (see reaction) 1). The two molecules of 3-phosphoglycerate–labelled "upper" and "lower" on the figure–are distinguished by the fact that the upper molecule contains the newly incorporated carbon dioxide, designated here as $^*CO_2$. Two properties of the carboxylase reaction are especially important:

- ✰ The negative change in free energy associated with the carboxylation of ribulose-1,5-bisphosphate is large; thus the forward reaction is strongly favored.
- ✰ The affinity of rubisco for CO_2 is sufficiently high to ensure rapid carboxylation at the low concentrations of CO_2 found in photosynthetic cells.

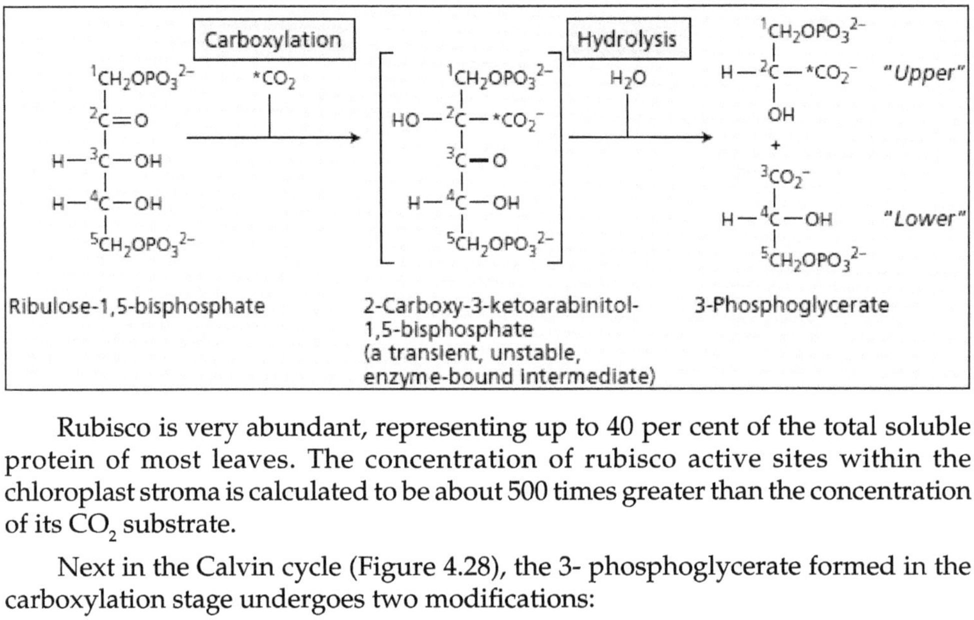

Rubisco is very abundant, representing up to 40 per cent of the total soluble protein of most leaves. The concentration of rubisco active sites within the chloroplast stroma is calculated to be about 500 times greater than the concentration of its CO_2 substrate.

Next in the Calvin cycle (Figure 4.28), the 3- phosphoglycerate formed in the carboxylation stage undergoes two modifications:

1. It is first phosphorylated via 3-phosphoglycerate kinase to 1,3-bisphosphoglycerate through use of the ATP generated in the light.
2. Then it is reduced to glyceraldehyde-3-phosphate through use of the NADPH generated by the light reactions The chloroplast enzyme NADP:glyceraldehyde-3-phosphate dehydrogenase catalyzes this step. Note that the enzyme is similar to that of glycolysis except that NADP rather than NAD is the coenzyme. An NADP-linked form of the enzyme is synthesized during chloroplast development (greening), and this form is preferentially used in biosynthetic reactions.

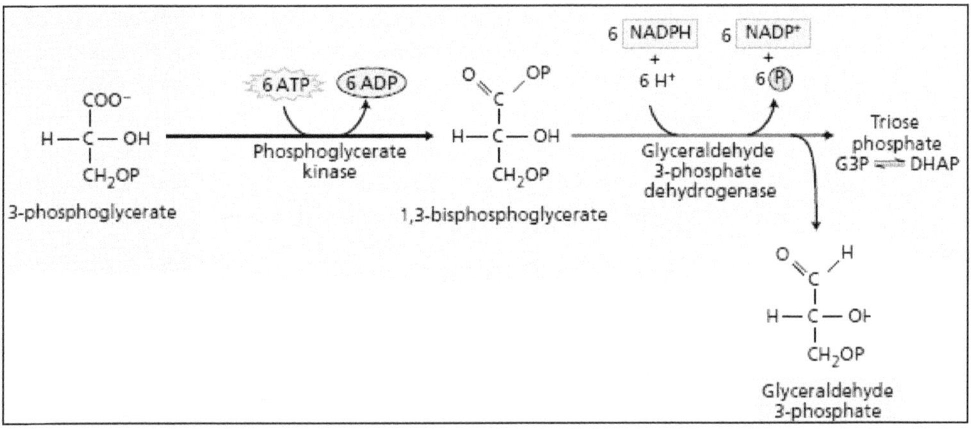

Regeneration of Ribulose-1,5-Bisphosphate

The continued uptake of CO_2 requires that the CO_2 acceptor, ribulose-1,5-bisphosphate, be constantly regenerated. To prevent depletion of Calvin cycle intermediates, three molecules of ribulose-1,5-bisphosphate (15 carbons total) are formed by reactions that reshuffle the carbons from the five molecules of triose phosphate (5 ??3 = 15 carbons).

1. One molecule of glyceraldehyde-3-phosphate is converted via triose phosphate isomerase to dihydroxyacetone- 3-phosphate in an isomerization reaction.
2. Dihydroxyacetone-3-phosphate then undergoes aldol condensation with a second molecule of glyceraldehyde- 3-phosphate, a reaction catalyzed by aldolase to give fructose-1,6-bisphosphate.
3. Fructose-1,6-bisphosphate occupies a key position in the cycle and is hydrolyzed to fructose-6-phosphate which then reacts with the enzyme transketolase.
4. A two-carbon unit (C-1 and C-2 of fructose-6-phosphate) is transferred via transketolase to a third molecule of glyceraldehyde-3-phosphate to give erythrose- 4-phosphate (from C-3 to C-6 of the fructose) and xylulose5-phosphate (from C-2 of the fructose and the glyceraldehyde-3-phosphate).
5. Erythrose-4-phosphate then combines via aldolase with a fourth molecule of triose phosphate (dihydroxyacetone- 3-phosphate) to yield the seven-carbon sugar sedoheptulose-1,7-bisphosphate.
6. This seven-carbon bisphosphate is then hydrolyzed by way of a specific phosphatase to give sedoheptulose-7-phosphate.
7. Sedoheptulose-7-phosphate donates a two-carbon unit to the fifth (and last) molecule of glyceraldehyde-3-phosphate via transketolase and produces ribose-5-phosphate (from C-3 to C-7 of sedoheptulose) and xylulose-5-phosphate (from C-2 of the sedoheptulose and the glyceraldehyde-3-phosphate).

8. The two molecules of xylulose-5-phosphate are converted to two molecules of ribulose-5-phosphate sugars by a ribulose-5-phosphate epimerase. The third molecule of ribulose-5-phosphate is formed from ribose-5-phosphate by ribose-5-phosphate isomerase.
9. Finally, ribulose-5-phosphate kinase catalyzes the phosphorylation of ribulose-5-phosphate with ATP, thus regenerating the three needed molecules of the initial CO_2 acceptor, ribulose-1,5-bisphosphate.

4.18.1. Regulation of the Calvin Cycle

Five light-regulated enzymes regulate the Calvin cycle:

1. Rubisco
2. NADP:glyceraldehyde-3-phosphate dehydrogenase
3. Fructose-1,6-bisphosphatase
4. Sedoheptulose-1,7-bisphosphatase
5. Ribulose-5-phosphate kinase

4.18.1.2. Regulation of Rubisco Activity

The activity of rubisco is also regulated by light, but the enzyme itself does not respond to thioredoxin. George Lorimer and colleagues found that rubisco is activated when activator CO_2 (a different molecule from the substrate CO_2 that becomes fixed) reacts slowly with an uncharged NH_2 group of lysine within the active site of the enzyme. The resulting carbamate derivative (a new anionic site) then rapidly binds Mg^{2+} to yield the activated complex (Figure 4.28). Two protons are released during the formation of the ternary complex rubisco–CO_2–Mg^{2+}, so activation is promoted by an increase in both pH and Mg^{2+} concentration. Thus, light-dependent stromal changes in pH and Mg^{2+} appear to facilitate the observed activation of rubisco by light. In the active state, rubisco binds another molecule of CO_2, which reacts with the 2,3-enediol form of ribulose- 1,5-bisphosphate (P–O–CH_2–COH—COH–CHOH– CH_2O–P) yielding 2-carboxy-3-ketoribitol 1,5-bisphos-phate. The extreme instability of the latter intermediate leads to the cleavage of the bond that links carbons 2 and 3 of ribulose-1,5-bisphosphate, and as a consequence, rubisco

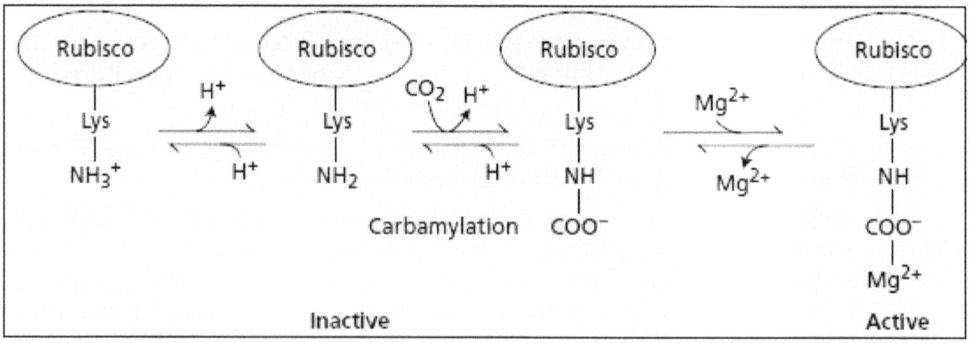

Figure 4.29b: Activation of Rubisco.

releases two molecules of 3-phosphoglycerate. The binding of sugar phosphates, such as ribulose-1,5- bisphosphate, to rubisco prevents carbamylation. The sugar phosphates can be removed by the enzyme rubisco activase, in a reaction that requires ATP. The primary role of rubisco activase is to accelerate the release of bound sugar phosphates, thus preparing rubisco for carbamylation. Rubisco is also regulated by a natural sugar phosphate, carboxyarabinitol-1-phosphate, that closely resembles the six-carbon transition intermediate of the carboxylation reaction. This inhibitor is present at low concentrations in leaves of many species and at high concentrations in leaves of legumes such as soybean and bean. Carboxyarabinitol-1-phosphate binds to rubisco at night, and it is removed by the action of rubisco activase in the morning, when photon flux density increases.

Recent work has shown that in some plants rubisco activase is regulated by the ferredoxin–thioredoxin system (Zhang and Portis 1999). In addition to connecting thioredoxin to all five regulatory enzymes of the Calvin cycle, this finding provides a new mechanism for linking light to the regulation of enzyme activity.

4.19. The C_2 or Oxidative Photosynthetic Carbon Cycle (Photorespiration)

An important property of rubisco is its ability to catalyze both the carboxylation and the oxygenation of RuBP. Oxygenation is the primary reaction in a process known as photorespiration. Because photosynthesis and photorespiration work in diametrically opposite directions, photorespiration results in loss of CO_2 from cells that are simultaneously fixing CO_2 by the Calvin cycle (Ogren 1984; Leegood *et al.*, 1995).

The incorporation of one molecule of O_2 into the 2,3-enediol isomer of ribulose-1,5-bisphosphate generates an unstable intermediate that rapidly splits into 2-phosphoglycolate and 3-phosphoglycerate (Figure 4.30).The ability to catalyze the oxygenation of ribulose- 1,5-bisphosphate is a property of all rubiscos. As alternative substrates for rubisco, CO_2 and O_2 compete for reaction with ribulose-1, 5-bisphosphate because carboxylation and oxygenation occur within the same active site of the enzyme. Angiosperm rubiscos fix CO_2 about 80 times faster than they oxygenate. However, an aqueous solution in equilibrium with air at 25°C has a CO_2:O_2 ratio of 0.0416. At these concentrations, carboxylation in air outruns oxygenation by a scant three to one. The C_2 oxidative photosynthetic carbon cycle acts as a scavenger operation to recover fixed carbon lost during photorespiration by the oxygenase reaction of rubisco. The 2-phosphoglycolate formed in the chloroplast by oxygenation of ribulose-1,5-bisphosphate is rapidly hydrolyzed to glycolate by a specific chloroplast phosphatase (Figure 4.30). Subsequent metabolism of the glycolate involves the cooperation of two other organelles: peroxisomes and mitochondria. Glycolate leaves the chloroplast via a specific transporter protein in the envelope membrane and diffuses to the peroxisome. There it is oxidized to glyoxylate and hydrogen peroxide (H_2O_2) by a flavin mononucleotide dependent oxidase: glycolate oxidase (Figure 4.30). The vast amounts of hydrogen peroxide released in the peroxisome are destroyed by the action of catalase while the glyoxylate undergoes transamination. The amino donor for this transamination

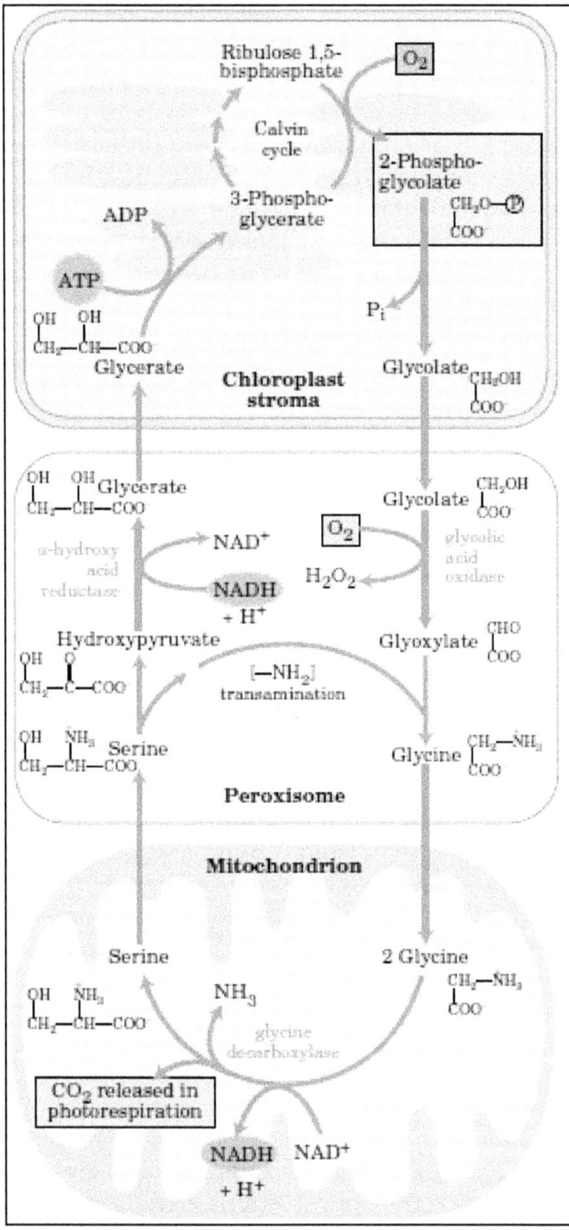

Figure 4.30: The Biochemical Pathway of Photorespiration.

is probably glutamate, and the product is the amino acid glycine. Glycine leaves the peroxisome and enters the mitochondrion (Figure 4.30). There the glycine decarboxylase multienzyme complex catalyzes the conversion of two molecules of glycine and one of NAD^+ to one molecule each of serine, NADH, NH_4^+ and CO_2. This multienzyme complex, present in large concentrations in the matrix of plant mitochondria, comprises four proteins, named H-protein (a lipoamide-containing polypeptide), P-protein (a 200 kDa, homodimer, pyridoxal phosphate-containing protein), T-protein (a folate-dependent protein), and L-protein (a flavin adenine nucleotide–containing protein). The ammonia formed in the oxidation of glycine diffuses rapidly from the matrix of mitochondria to chloroplasts, where glutamine synthetase combines it with carbon skeletons to form amino acids. The newly formed serine leaves the mitochondria and enters the peroxisome, where it is converted first by transamination to hydroxypyruvate and then by an NADH dependent reduction to glycerate. Amalate-oxaloacetate shuttle transfers NADH from the cytoplasm into the peroxisome, thus maintaining an adequate concentration of NADH for this reaction. Finally, glycerate reenters the chloroplast, where it is phosphorylated to yield 3-phosphoglycerate. In photorespiration, various compounds are circulated in concert through two cycles. In one of the cycles, carbon exits the chloroplast in two molecules of glycolate and returns in one molecule of glycerate. In the other cycle, nitrogen exits the chloroplast in one molecule of glutamate and returns in one molecule of ammonia (together with one molecule of α-ketoglutarate) (see Figure 4.29b). Thus overall, two molecules of phosphoglycolate (four carbon atoms), lost from the Calvin cycle by the oxygenation of RuBP, are converted into one molecule of 3-phosphoglycerate (three carbon atoms) and one CO_2. In other words, 75 per cent of the carbon lost by the oxygenation of ribulose-1, 5-bisphosphateis recovered by the C_2 oxidative photosynthetic carbon cycle and returned to the Calvin cycle (Lorimer 1981). On the other hand, the total organic nitrogen remains unchanged because the formation of inorganic nitrogen (NH_4^+) in the mitochondrion is balanced by the synthesis of glutamine in the chloroplast. Similarly, the use of NADH in the peroxisome (by hydroxypyruvate reductase) is balanced by the reduction of NAD^+ in the mitochondrion.

4.20. Competition between Carboxylation and Oxygenation Decreases the Efficiency of Photosynthesis

Because photorespiration is concurrent with photosynthesis, it is difficult to measure the rate of photorespiration in intact cells. Two molecules of 2-phosphoglycolate (four carbon atoms) are needed to make one molecule of 3-phosphoglycerate, with the release of one molecule of CO_2; so theoretically one-fourth of the carbon entering the C_2 oxidative photosynthetic carbon cycle is released as CO_2. Measurements of CO_2 release by sunflower leaves support this calculated value. This result indicates that the actual rate of photosynthesis is approximately 120 to 125 per cent of the measured rate. The ratio of carboxylation to oxygenation in air at 25°C is computed to be between 2.5 and 3. Further calculations indicate that photorespiration lowers the efficiency of photosynthetic carbon fixation from 90 per cent to approximately 50 per cent. This decreased efficiency can be measured as an increase in the quantum requirement for CO_2 fixation under photorespiratory conditions (air with high O_2 and low CO_2) as opposed to nonphotorespiratory conditions (low O_2 and high CO_2).

4.21. CO_2-Concentration Mechanisms

Many plants either do not photorespire at all, or they do so to only a limited extent. These plants have normal rubiscos, and their lack of photorespiration is a consequence of mechanisms that concentrate CO_2 in the rubisco environment and thereby suppress the oxygenation reaction. Three mechanisms for concentrating CO_2 at the site of carboxylation:

1. C_4 photosynthetic carbon fixation (C_4)
2. Crassulacean acid metabolism (CAM)
3. CO_2 pumps at the plasma membrane

4.21.1. The C_4 Carbon Cycle (Hatch and Slack Cycle)

There are differences in leaf anatomy between plants that have a C_4 carbon cycle (called *C_4 plants*) and those that photosynthesize solely via the Calvin photosynthetic cycle (*C_3 plants*). Across section of a typical C_3 leaf reveals one major cell type that has chloroplasts, the mesophyll. In contrast, a typical C_4 leaf has two distinct chloroplast-containing cell types: mesophyll and bundle sheath (or *Kranz*, German for "wreath")

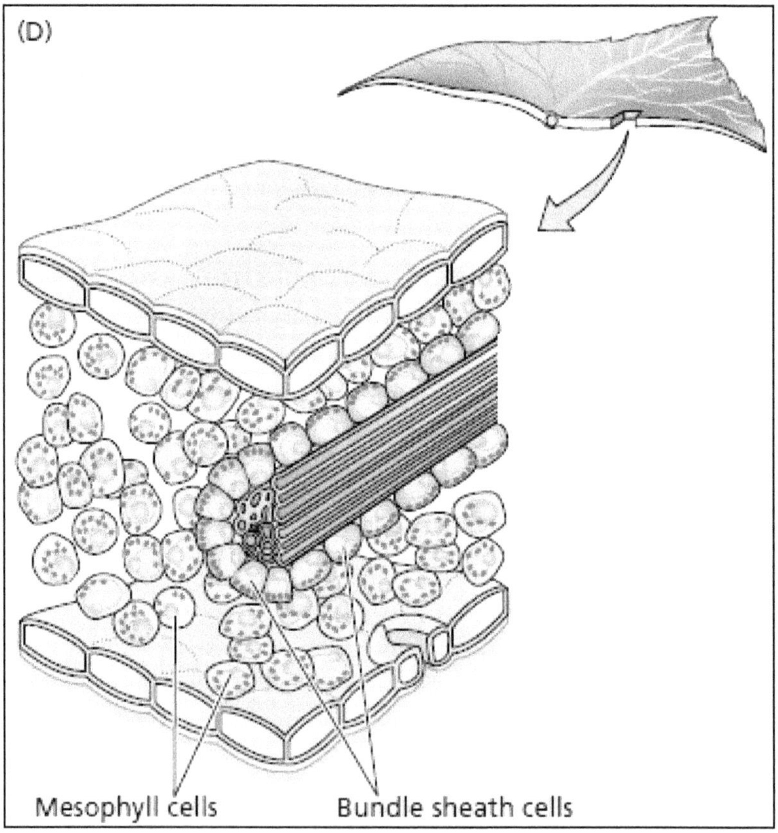

Figure 4.31: Anatomical Feature of C_4 Leaf.

cells (Figure 2.4). The chloroplast in C_4 leaves are dimorphic (two morphologically distinct types).The chloroplast of bundle sheath cells are larger in size and arranged centripetally. They contain starch grains but lack grana. The mesophyll cells, on the other hand, contain normal type of chloroplast. The mesophyll cells in C_4 plants perform C_4 cycle and the cells of bundle sheath perform C_3 cycle. In the tropical grasses, sugarcane, and maize, the C4 cycle is now known to occur in 16 families of both monocotyledons and dicotyledons, and it is particularly prominent in Gramineae (corn, millet, sorghum, sugarcane), Chenopodiaceae (*Atriplex*), and Cyperaceae (sedges). About 1 per cent of all known species have $_{C4}$ metabolism. There is considerable anatomic variation in the arrangement of the bundle sheath cells with respect to the mesophyll and vascular tissue. In all cases, however, operation

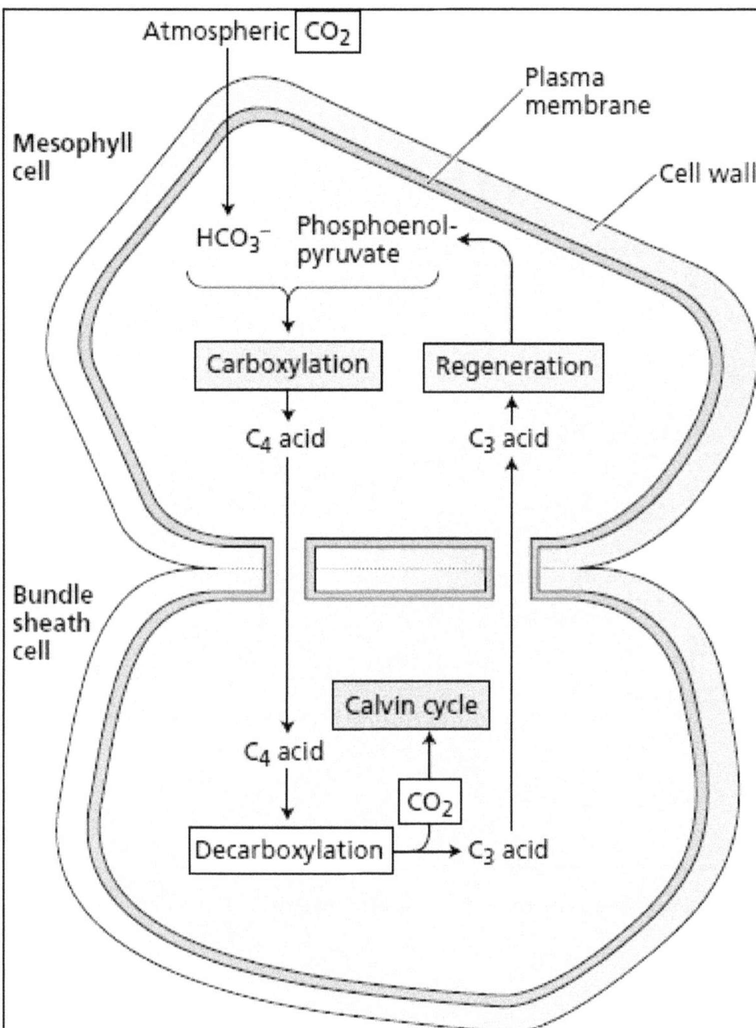

Figure 4.32

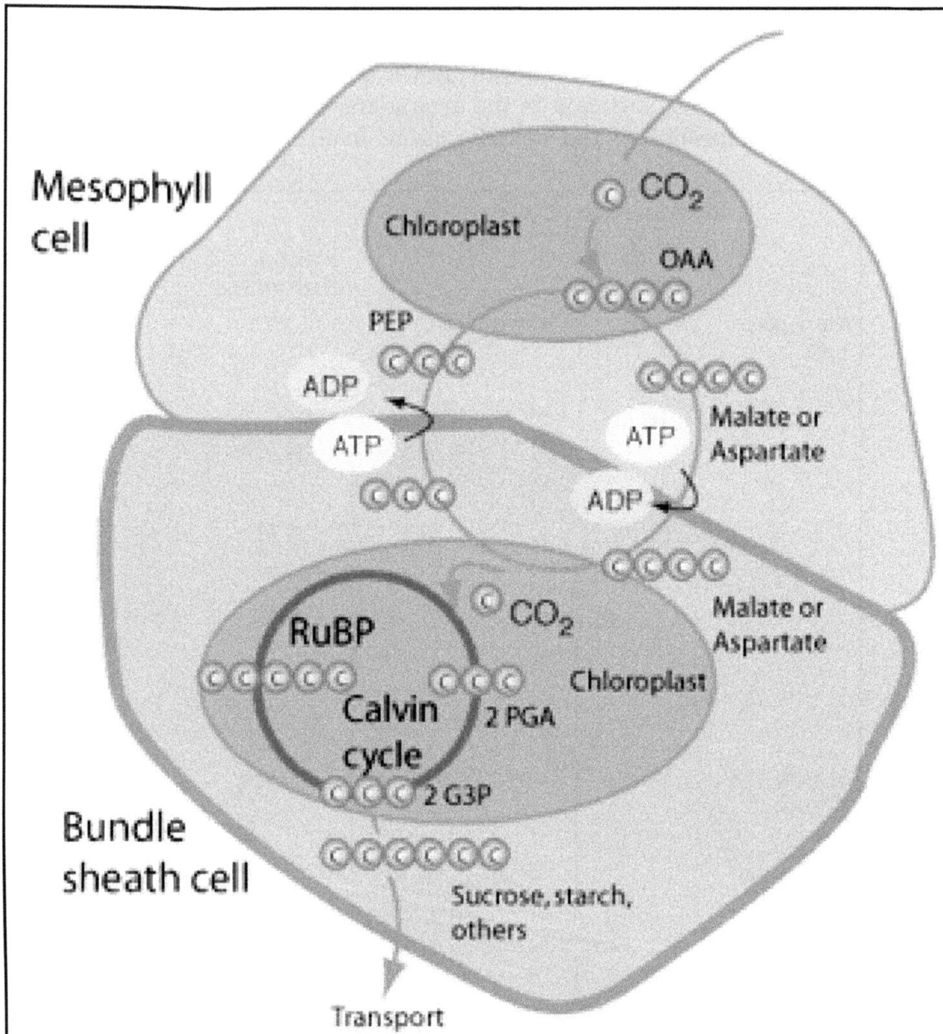

Figure 4.33: The Diagrammatic Representation of Hatch and Slack Cycle.

of the C_4 cycle requires the cooperative effort of both cell types. No mesophyll cell of a C_4 plant is more than two or three cells away from the nearest bundle sheath cell (Figure 4.31). In addition, an extensive network of plasmodesmata connects mesophyll and bundle sheath cells, thus providing a pathway for the flow of metabolites between the cell types.

The basic C_4 cycle consists of four stages:

1. Fixation of CO_2 by the carboxylation of phosphoenolpyruvate in the mesophyll cells to form a C_4 acid (malate and/or aspartate)
2. Transport of the C_4 acids to the bundle sheath cells
3. Decarboxylation of the C_4 acids within the bundle sheath cells and generation of CO_2, which is then reduced to carbohydrate via the Calvin cycle.
4. Transport of the C_3 acid (pyruvate or alanine) that is formed by the decarboxylation step back to the mesophyll cell and regeneration of the CO_2 acceptor phosphoenolpyruvate.

One interesting feature of the cycle is that regeneration of the primary acceptor–phosphoenolpyruvate–consumes two "high-energy" phosphate bonds: one in the reaction catalyzed by pyruvate–orthophosphate dikinase (reaction 8) and another in the conversion of PPi to 2Pi catalyzed by pyrophosphatase. The leaves of C_4 plants require at least 5 ATP and 2 mol. of reduced NADP to fix 1 mol. of CO_2

4.21.2. Malate and Aspartate are Carboxylation Products of the C_4 Cycle

Early labeling of C_4 acids was first observed in $14CO_2$ labeling studies of sugarcane by H. P. Kortschack and colleagues and of maize by Y. Karpilov and coworkers. When leaves were exposed for a few seconds to $14CO_2$ in the light, 70 to 80 per cent of the label was found in the C_4 acids malate and aspartate–a pattern very different from the one observed in leaves that photosynthesize solely via the Calvin cycle. In pursuing these initial observations, M. D. Hatch and C. R. Slack elucidated what is now known as the C_4 photosynthetic carbon cycle (C_4 cycle) (Figure 4.34). They established that the C_4 acids malate and aspartate are the first stable, detectable intermediates of photosynthesis in leaves of sugarcane and that carbon atom 4 of malate subsequently becomes carbon atom 1 of 3-phosphoglycerate (Hatch and Slack 1966). The primary carboxylation in these leaves is catalyzed not by rubisco, but by PEP (phosphoenylpyruvate) carboxylase.

Ecological Significance of C_4 (Efficiency of C_4 plants compared to C_3 plants)

Unlike C_3 plants, photosynthesis of C_4 plants is not inhibited by O_2, and they exhibit no postillumination CO_2 burst and have a very low CO_2 concentration (at which the rate of CO_2 uptake is balanced by the rate of CO_2 evolution).

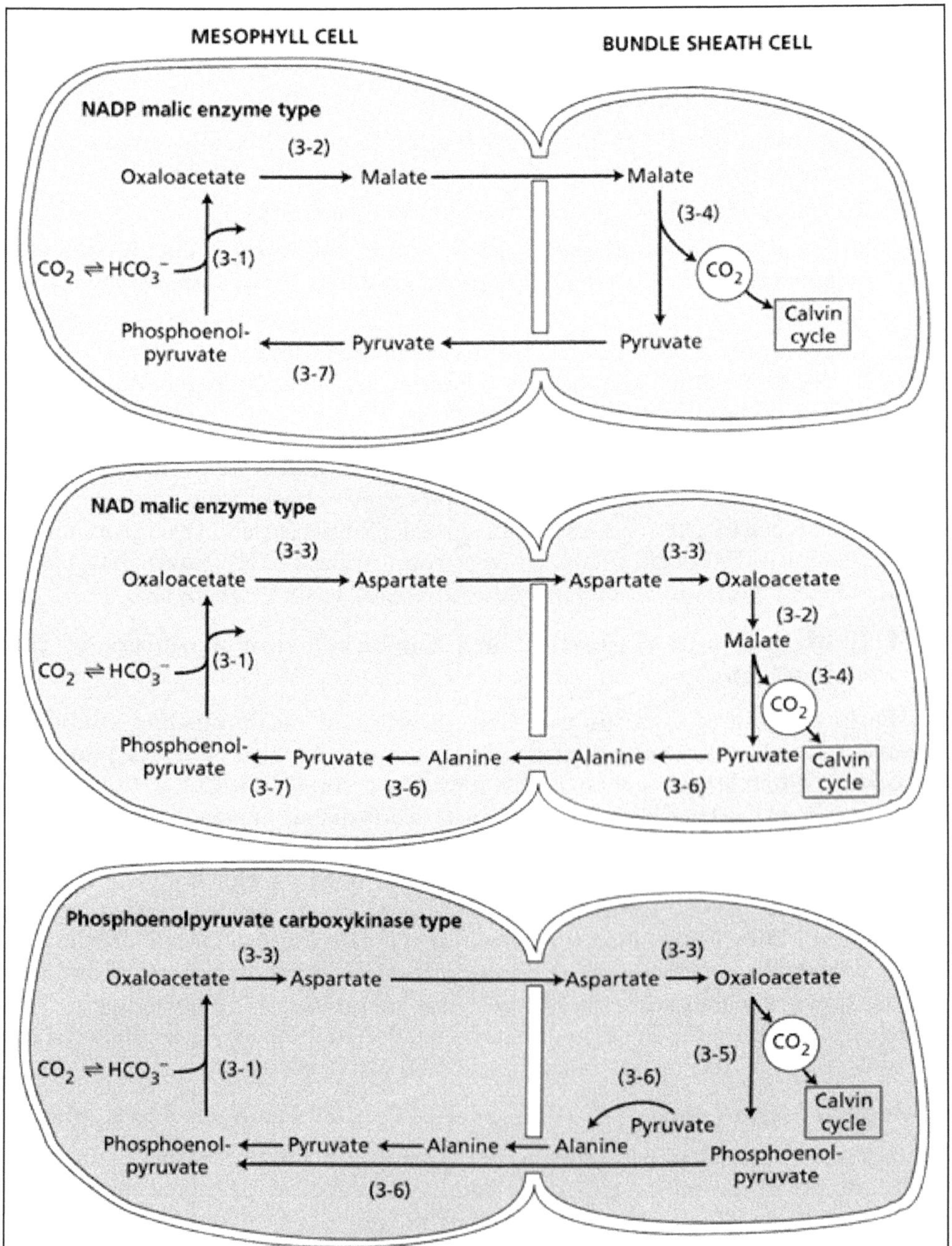

Figure 4.34: Three Variation of the C_4 Cycle (Courtsey of Hopkins).

A Comparison of Significant Feature of C_3 and C_4 Plants

	C_3	C_4
Photorespiration	Yes	No
CO_2 compensation point (µl CO_2 l^{-1})	20-100	0-5
Temperature optimum (ºC)		
Photosynthesis	20-25	30-45
Rubisco	20-25	—
PEPcase	—	30-35
Quantum yield as a function of temperature	Declining	Steady
Transpiration ratio	500-1000	200-350

The high level of CO_2 level developed in bundle sheath cell would tend to suppress photorespiration by out competing O_2 for binding to Rubisco. In addition, the anatomical and biochemical adaptations of C4 leaves ensure that any CO_2 might escape the bundle sheath cell is trapped and reassimilated by PEPcase in mesophyll cells,before it has the opportunity to escape from the leaf.Thus C4 leaves are not only efficient CO_2 absorbers, but also effectively trap and recirculate any CO_2 that might be produced in the leaf. The quantum yield of photosynthesis in C_3 plants tends to decline with increasing leaf temperature while the quantum yields of C_4 plants remain essentially constant. The decline of quantum yield in C_3 plants is due to decreased carboxylation activity of rubisco at higher temperature. The solubility of gases generally declines with increasing temperature,but solubility of CO_2 is affected more than the solubility of O_2. So higher temperature favour oxygenation by rubisco. C_4 plants can maintain higher rates of photosynthesis at lower CO_2 levels. Thus C_4 plants gain an advantages over C_3 plants when stomata are partially closed to conserve water during water stress.

4.22. Crassulacean Acid Meteabolism (CAM)

A third mechanism for concentrating CO_2 at the site of rubisco is found in crassulacean acid metabolism (CAM). Despite its name, CAM is not restricted to the family Crassulaceae (*Crassula, Kalanchoe, Sedum*); it is found in numerous angiosperm families. Cacti and euphorbias are CAM plants, as well as pineapple, vanilla, and agave. The CAM mechanism enables plants to improve water use efficiency. Typically, CAM plant loses 50 to 100 g of water for every gram of CO_2 gained, compared with values of 250 to 300 g and 400 to 500 g for C_4 and C_3 plants, respectively. Thus, CAM plants have a competitive advantage in dry environments. The CAM mechanism is similar in many respects to the C4 cycle. In C_4 plants, formation of the C4 acids in the mesophyll is spatially separated from decarboxylation of the C_4 acids and from refixation of the resulting CO_2 by the Calvin cycle in the bundle sheath. In CAM plants, formation of the C_4 acids is both temporally and spatially separated. At night, CO_2 is captured by PEP carboxylase in the cytosol, and the malate that forms from the oxaloacetate product is stored in the vacuole (Figure 2.8). During the day, the stored malate is transported to the

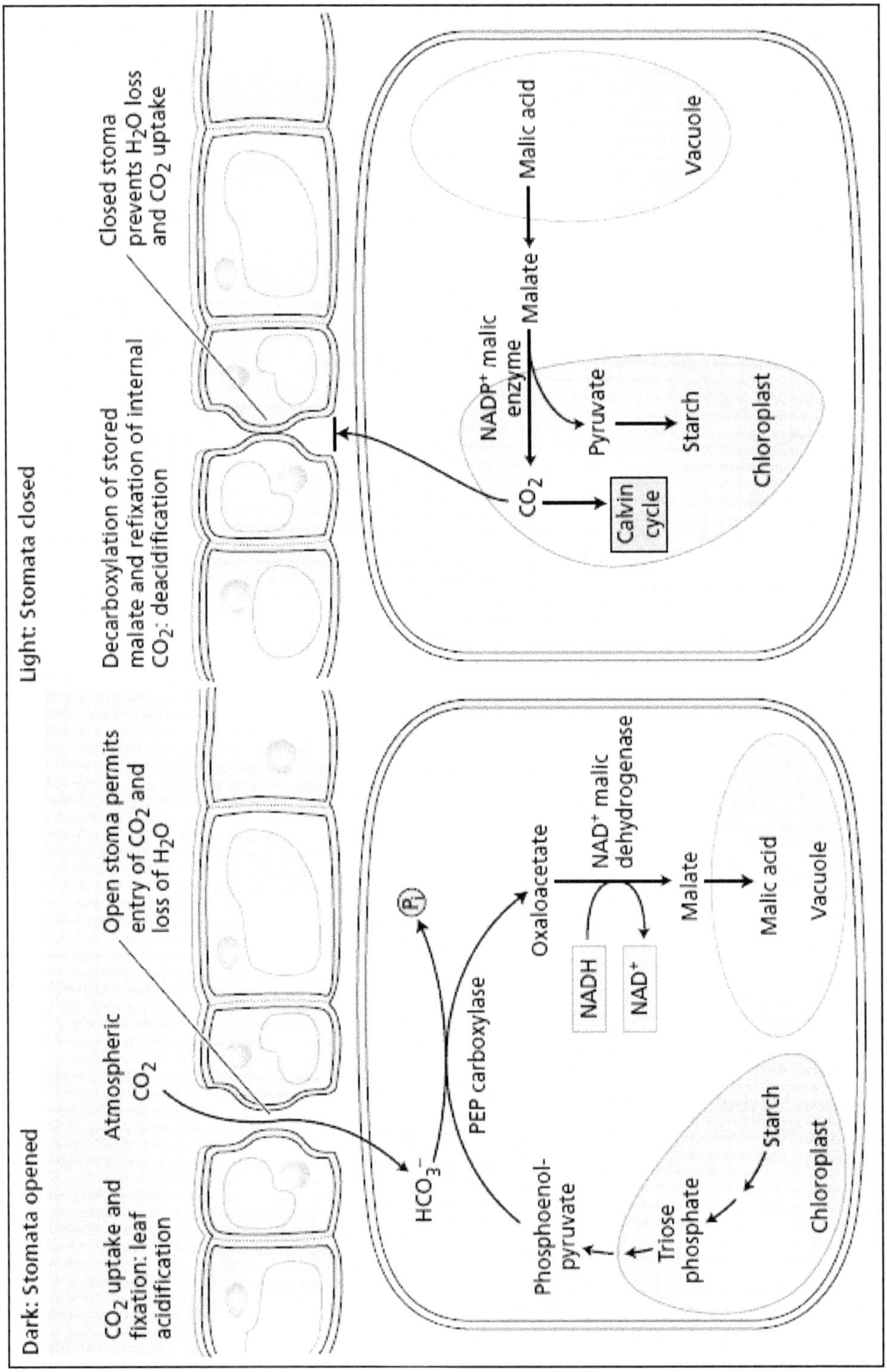

Figure 4.35: CAM Pathway.

chloroplast and decarboxylated by NADP-malic enzyme, the released CO_2 is fixed by the Calvin cycle, and the NADPH is used for converting the decarboxylated triose phosphate product to starch. CAM plants such as cacti achieve their high water use efficiency by opening their stomata during the cool, desert nights and closing them during the hot, dry days. Closing the stomata during the day minimizes water loss, but because H_2O and CO_2 share the same diffusion pathway, CO_2 must then be taken up at night.

4.23. Algal and Cyanobacterial CO_2 Pumps

When algal and cyanobacterial cells are grown in air enriched with 5 per cent CO_2 and then transferred to a low-CO_2 medium, they display symptoms typical of photorespiration (O_2 inhibition of photosynthesis at low concentration of CO_2). But if the cells are grown in air containing 0.03 per cent CO_2, they rapidly develop the ability to concentrate inorganic carbon (CO_2 plus HCO_3 –) internally. Under these low-CO_2 conditions, the cells no longer photorespire. At the concentrations of CO_2 found in aquatic environments, rubisco operates far below its maximal specific activity. Marine and freshwater organisms overcome this drawback by accumulating inorganic carbon by the use of CO_2 and HCO_3 – pumps at the plasma membrane. ATP derived from the light reactions provides the energy necessary for the active uptake of CO_2 and HCO_3 –.Recent work indicates that a single gene encoding a transcription factor can regulate the expression of genes that encode the components of the CO_2-concentrating mechanism in algae (Xiang *et al.*, 2001). The proteins that function as CO_2–HCO_3 – pumps are not present in cells grown in high concentrations of CO_2 but are induced upon exposure to low concentrations of CO_2. The accumulated HCO_3 – is converted to CO_2 by the enzyme carbonic anhydrase, and the CO_2 enters the Calvin cycle. The metabolic consequence of this CO_2 enrichment is suppression of the oxygenation of ribulose bisphosphate and hence also suppression of photorespiration. The energetic cost of this adaptation is the additional ATP needed for concentrating the CO_2.

Facter affecting the Photosynthesis

F. F. Blackman hypothesized in 1905 that, under any particular conditions, the rate of photosynthesis is limited by the slowest step, the so-called *limiting factor*. The implication of this hypothesis is that at any given time, photosynthesis can be limited either by light or by CO_2 concentration, but not by both factors.

4.24. Chloroplast Movement and Leaf Movement can control Light Absorption

Chloroplast movement is widespread among algae, mosses, and leaves of higher plants (Haupt and Scheuerlein 1990). If chloroplast orientation and location are controlled, leaves can regulate how much of the incident light is absorbed. Under low light, chloroplasts gather at the cell surfaces parallel to the plane of the leaf so that they are aligned perpendicularly to the incident light– a position that maximizes absorption of light. Under high light, the chloroplasts move to the cell surfaces that are parallel to the incident light, thus avoiding excess absorption of light. Such chloroplast rearrangement can decrease the amount of light absorbed by

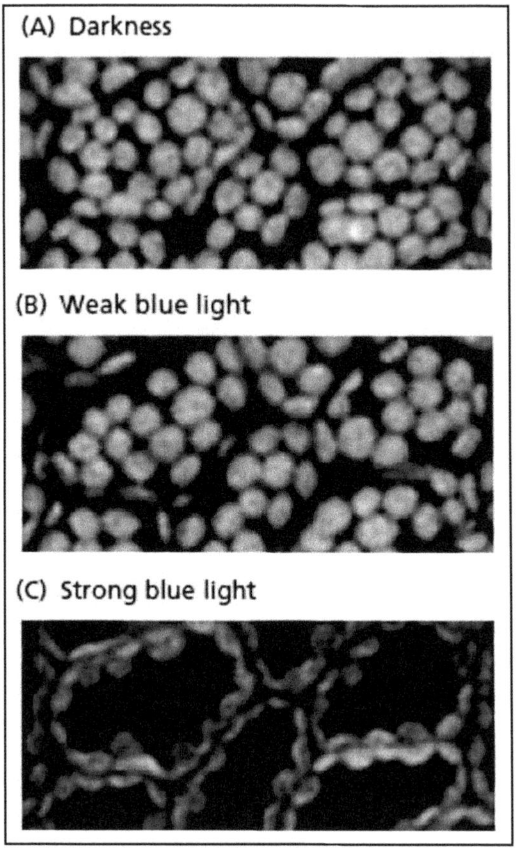

the leaf by about 15 per cent (Gorton *et al.*, 1999). Chloroplast movement in leaves is a typical blue-light response. Blue light also controls chloroplast orientation in many of the lower plants, but in some algae, chloroplast movement is controlled by phytochrome.

4.25. Photosynthetic Responses to Light

Measuring CO_2 fixation in intact leaves at increasing photon flux allows us to construct light-response curves that provide useful information about the photosynthetic properties of leaves. In the dark there is no photosynthetic carbon assimilation, and CO_2 is given off by the plant because of respiration. By convention CO_2 assimilation is negative in this part of the light-response curve. As the photon flux increases, photosynthetic CO_2 assimilation increases until it equals CO_2 release by mitochondrial respiration. The point at which CO_2 uptake exactly balances CO_2 release is called the light compensation point. The photon flux at which different leaves reach the light compensation point varies with species and developmental conditions. One of the more interesting differences is found between plants grown in full sunlight and those grown in the shade (Figure 4.35). Light compensation points of sun plants range from 10 to 20 μmol m^{-2} s^{-1}; corresponding values for shade plants are 1 to 5 mol m–2 s–1. The values for shade plants are lower because respiration rates in shade plants are very low, so little net photosynthesis suffices to bring the net rates of CO_2 exchange to zero. Low respiratory rates seem to represent a basic adaptation that allows shade plants to survive in light-limited environments. Increasing photon flux above the light compensation point results in a proportional increase in photosynthetic rate (Figure 4.35), yielding a linear relationship between photon flux and photosynthetic rate. Such a linear rela- tionship comes about because photosynthesis is *light limited* at those levels of incident light, so more light stimulates more photosynthesis. In this linear portion of the curve, the slope of the line reveals the maximum quantum yield of photosynthesis for the leaf. At higher photon fluxes, the photosynthetic response to light starts to level off (Figure 4.35) and reaches *saturation*. Once the saturation point is reached, further increases in photon flux no longer affect photosynthetic rates, indicating that factors other than incident light, such as electron transport rate, rubisco activity, or the metabolism of triose phosphates, have become limiting to photosynthesis. After the saturation point, photosynthesis is commonly referred to as CO_2 limited, reflecting the inability of the Calvin cycle enzymes to keep pace with the absorbed light energy.

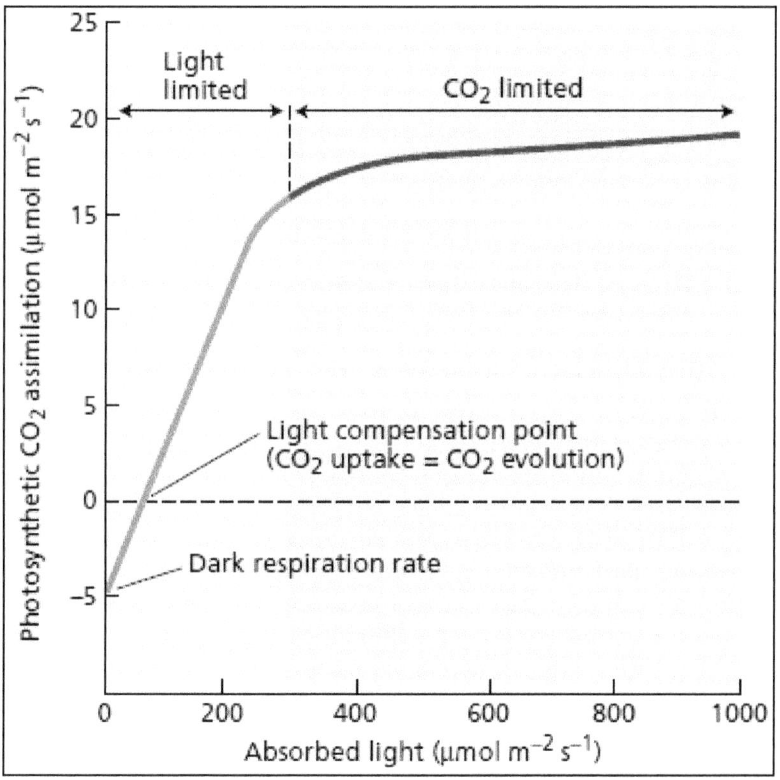

Figure 4.36: Response of Photosynthesis to Light in C_3 Plants (Courtesy of Taiz and Zieger).

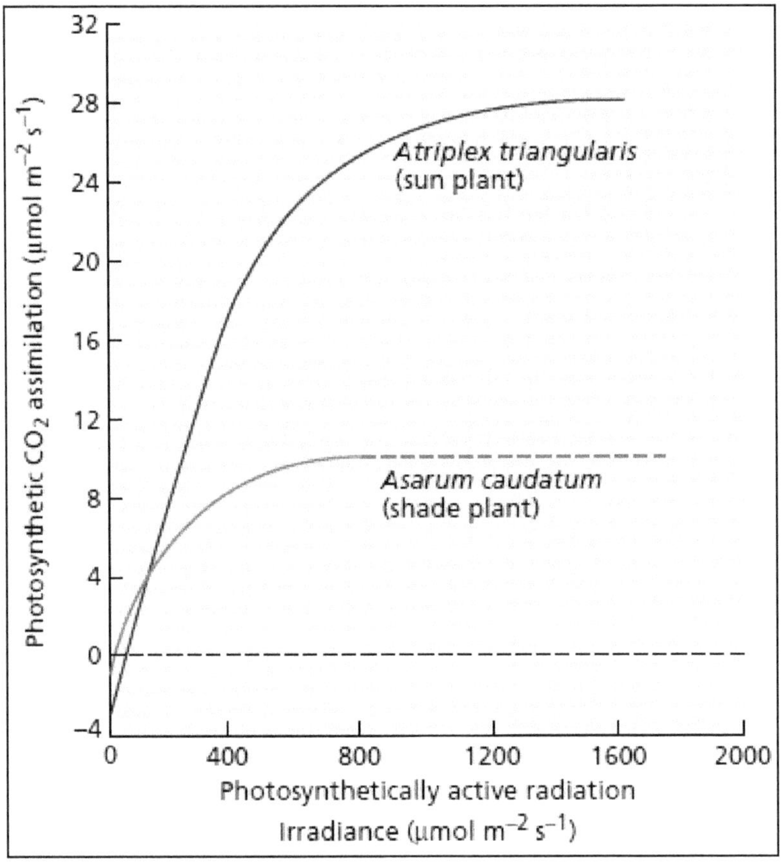

Figure 4.37: Light Response Curves of Photosynthetic Carbon Fixation in Sun and Shade Plants (Courtesy of Taiz and Zieger).

4.26. Photosynthetic Responses to Carbon Dioxide

CO_2 diffuses from the atmosphere into leaves–first through stomata, then through the intercellular air spaces, and ultimately into cells and chloroplasts. In the presence of adequate amounts of light, higher CO_2 concentrations support higher photosynthetic rates. The reverse is also true; that is, low CO_2 concentration can limit the amount of photosynthesis. We consider the limitations that CO_2 places on photosynthesis and the impact of the CO_2-concentrating mechanisms of C_4 plants. Carbon dioxide is a trace gas in the atmosphere, presently accounting for about 0.037 per cent, or 370 parts per million (ppm), of air. The partial pressure of ambient CO_2 (Ca) varies with atmospheric pressure and is approximately 36 pascals (Pa) at sea level. The current CO_2 concentration of the atmosphere is increasing by about 1 ppm each year, primarily because of the burning of fossil fuels and by 2020 the atmospheric CO_2 concentration could reach 600 ppm. The increased CO_2 concentration and temperature associated with the greenhouse effect can influence photosynthesis. At current atmospheric CO_2 concentrations, photosynthesis in C_3 plants is CO_2 limited, but this situation could change as atmospheric CO_2

concentrations continue to rise. Under laboratory conditions, most C_3 plants grow 30 to 60 per cent faster when CO_2 concentration is doubled (to 600–700 ppm), and the growth rate changes depend on nutrient status (Bowes, 1993). In some plants the enhanced growth is only temporary. For many crops, such as tomatoes, lettuce, cucumbers, and roses growing in greenhouses under optimal nutrition, carbon dioxide enrichment in the greenhouse environment results in increased productivity. The photosynthetic performance of C_3 plants under elevated CO_2 is enhanced because photorespiration decreases.

Expressing photosynthetic rate as a function of the partial pressure of CO_2 in the intercellular air space (Ci) within the leaf makes it possible to evaluate limitations to photosynthesis imposed by CO_2 supply. At very low intercellular CO_2 concentrations, photosynthesis is strongly limited by the low CO_2, while respiratory rates are unaffected. As a result, there is a negative balance between CO_2 fixed by photosynthesis and CO_2 produced by respiration, and a net efflux of CO_2 from the plant. Increasing intercellular CO_2 to the concentration at which these two processes balance each other defines the CO_2 compensation point, at which the net efflux of CO_2 from the plant is zero (Figure 2.11 A). This concept is analogous to that of the light compensation point discussed earlier in the chapter: The CO_2 compensation point reflects the balance between photosynthesis and respiration as a function of CO_2 concentration, and the light compensation point reflects that balance as a function of photon flux. In C_3 plants, increasing CO_2 above the compensation point stimulates photosynthesis over a wide concentration. range (see Figure 4.37a). At low to intermediate CO_2 concentrations, photosynthesis is limited by the carboxylation capacity of rubisco. At high CO_2 concentrations, photosynthesis is limited by the capacity of Calvin cycle to regenerate the acceptor

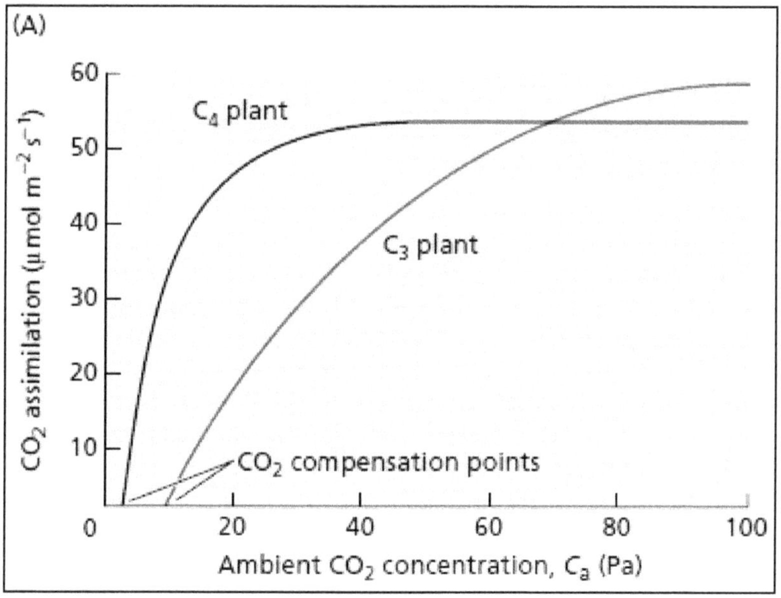

Figure 4.37a: Changes in Photosynthesis as a Function of Intercellular CO_2 Concentration (Courtesy of Taiz and Zieger).

molecule ribulose-1,5-bisphosphate, which depends on electron transport rates. By regulating stomatal conductance, most leaves appear to regulate their Ci (internal partial pressure for CO_2) such that it is intermediate between limitations imposed by carboxylation capacity and the capacity to regenerate ribulose-1,5-bisphosphate. A plot of CO_2 assimilation as a function intercellular partial pressures of CO_2 tells us how photosynthesis is regulated by CO_2, independent of the functioning of stomata (Figure 4.37b).

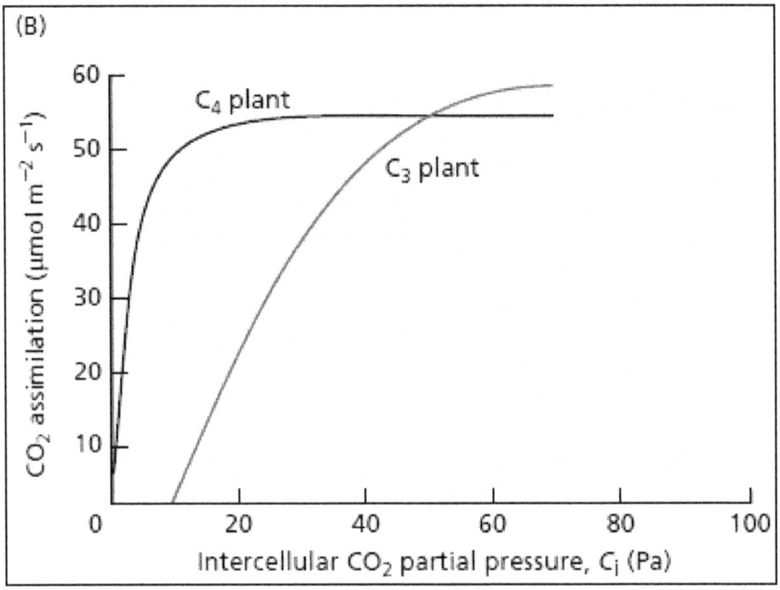

Figure 4.37b: Changes in Photosynthesis as a Function of Intercellular CO_2 Concentration.

4.27. Discrimination of Carbon Isotopes Reveals Different Photosynthetic Pathways

Atmospheric CO_2 contains the naturally occurring carbon isotopes 12C, 13C, and 14C in the proportions 98.9 per cent, 1.1 per cent, and 10–10 per cent, respectively. $14CO_2$ is present in such small quantities that it has no physiological relevance, but $13CO_2$ is different. The chemical properties of $13CO_2$ are identical to those of $12CO_2$, but because of the slight difference in mass (2.3 per cent), most plants assimilate less $13CO_2$ than $12CO_2$. In other words, plants discriminate against the heavier isotope of carbon, and they have smaller ratios of 13C to 12C than are found in atmospheric CO_2. Carbon isotope composition is measured by use of a mass spectrometer, which yields the following ratio:

$$R = \frac{^{13}CO_2}{^{12}CO_2}$$

4.28. Photosynthetic Responses to Temperature

When photosynthetic rate is plotted as a function of temperature, the curve has a characteristic bell shape (Figure 4.38). The ascending arm of the curve represents a temperature- dependent stimulation of photosynthesis up to an optimum; the descending arm is associated with deleterious effects, some of which are reversible while others are not. Temperature affects all biochemical reactions of photosynthesis, so it is not surprising that the responses to temperature are complex. We can gain insight into the underlying mechanisms by comparing photosynthetic rates in air at normal and at high CO_2 concentrations. At high CO_2 (Figure 4.39), there is an ample supply of CO_2 at the card 13C0 boxylation sites, and the rate of photosynthesis is limited primarily by biochemical reactions connected with electron transport. In these conditions, temperature changes have large effects on fixation rates. At ambient CO_2 concentrations (Figure 4.39), photosynthesis is limited by the activity of rubisco, and the response reflects two conflicting processes: an increase in carboxylation rate with temperature and a decrease in the affinity of rubisco for CO_2 as the temperature

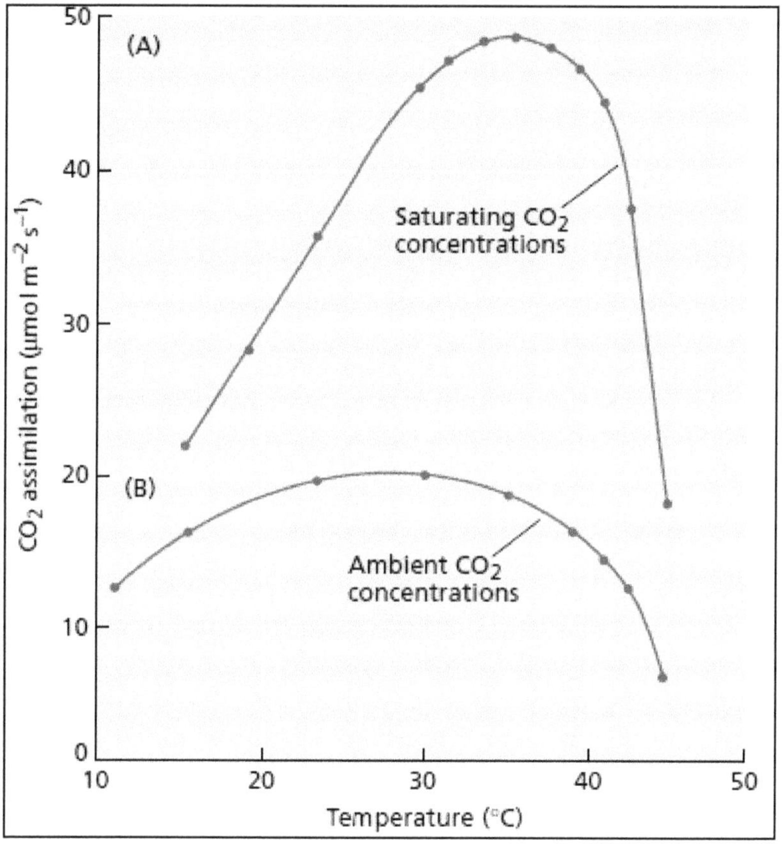

Figure 4.38: Channges in Photosynthesis as a Function of Temperature at CO_2 Concentration that Saturate Photosynthetic CO_2 Assimilation. (Courtesy of Taiz and Zieger).

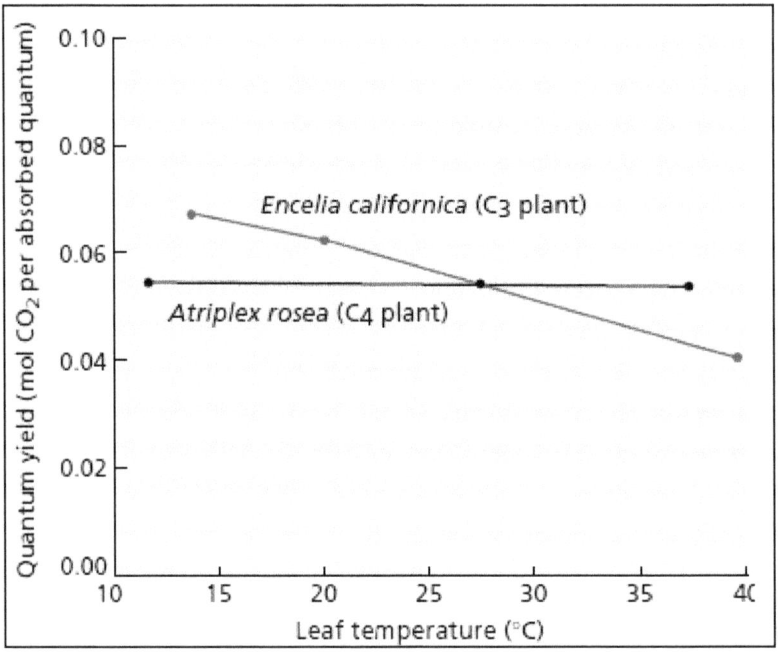

Figure 4.39: The Quantum Yield of Photosynthetic Carbon Fixation in C_3 and C_4 Plants (Courtesy of Taiz and Zieger).

rises. These opposing effects dampen the temperature response of photosynthesis at ambient CO_2 concentrations. Respiration rates also increase as a function of temperature, and the interaction between photorespiration and photosynthesis becomes apparent in temperature responses. Figure 4.39 shows changes in quantum yield as a function of temperature in a C_3 plant and in a C_4 plant. In the C_4 plant the quantum yield remains constant with temperature, reflecting typical low rates of photorespiration. In the C_3 plant the quantum yield decreases with temperature, reflecting a stimulation of photorespiration by temperature and an ensuing higher energy demand per net CO_2 fixed. At low temperatures, photosynthesis is often limited by phosphate availability at the chloroplast.

References

Bassi, R., Pineau, B., Dainese, P.and Marquardt, J . (1993). Carotenoid-binding proteins of photosystem II. *European Journal of Biochemistry*, 212: 297–303.

Elrad, D., Niyogi, K.K.and Grossman, A.R. (2002). A major light-harvesting polypeptide of photosystem II functions in thermal dissipation. *Plant Cell*, 141: 801–1816 .

Green, B.R. and Durnford, D.G. (1996). The chlorophyll-carotenoid proteins of oxygenic photosynthesis. *Plant Molecular Biology*, 47: 685–714.

Chapter 5

Respiration

Cellular respiration consists of a series of pathways by which carbohydrate and other molecules are oxidized for the purpose of retrieving the energy stored in photosynthesis and to obtain carbon compound used in the growth and maintenance of the cell. Higher plants are aerobic organisms, which means they require the presence of molecular oxygen for normal metabolism. They obtain both energy and carbon by oxidizing photoassimilates acoording to the following overall equation:

From chemical reaction, plant respiration can be expressed as the oxidation of the hexose to carbon di-oxide, with water as the product. Respiration is accompanied by the release of an amount of free energy.

$$C_6H_{12}O_6 + 6O_2 + 6H_2O \longrightarrow 6\,CO_2 + 12\,H_2O + \Delta G = -\,2872 \text{ kj mol}^{-1}$$

Respiratory Quotient (RQ)

R.Q. is the ratio of the volume of the carbon dioxide released to the volume of oxygen taken in respiration in the given period of time at standard temperature and pressure. The value of R.Q. varies with the respiratory substrates and their oxidation.

$$\text{R.Q.} = \frac{\text{Volume of } CO_2 \text{ evolved}}{\text{Volume of } O_2 \text{ absorbed}}$$

R.Q. of carbohydrate,

The respiration reactions can be grouped into four major processes:

1. Glycolysis,
2. Citric acid cycle,
3. Pentose phosphate pathway, and
4. Oxidative phosphorylation

5.1 Glycolysis

In the early steps of glycolysis (from the Greek words *glykos*, "sugar," and *lysis*, "splitting"), carbohydrates are converted to hexose phosphates, which are then split into two triose phosphates. In a subsequent energy-conserving phase, the triose phosphates are oxidized and rearranged to yield two molecules of pyruvate, an organic acid. Besides preparing the substrate for oxidation in the citric acid cycle, glycolysis yields a small amount of chemical energy in the form of ATP and NADH. When molecular oxygen is unavailable–for example, in plant roots in flooded soils–glycolysis can be the main source of energy for cells. For this to work, the fermentation pathways, which are localized in the cytosol, reduce pyruvate to recycle the NADH produced by glycolysis.

Plant glycolysis has unique regulatory features, as well as a parallel partial glycolytic pathway in plastids and alternative enzymatic routes for several cytosolic steps. In animals the substrate of glycolysis is glucose and the end product pyruvate. Because sucrose is the major translocated sugar in most plants and is therefore the form of carbon that most nonphotosynthetic tissues import, sucrose (not glucose) can be argued to be the true sugar substrate for plant respiration. The end products of plant glycolysis include another organic acid, malate. In the early steps of glycolysis, sucrose is broken down into the two monosaccharides–glucose and fructose– which can readily enter the glycolytic pathway. Two pathways for the degradation of sucrose are known in plants, both of which also take part in the unloading of sucrose from the phloem.

In the initial phase of glycolysis, each hexose unit is phosphorylated twice and then split, eventually producing two molecules of triose phosphate. This series of reactions consumes two to four molecules of ATP per sucrose unit, depending on whether the sucrose is split by sucrose synthase or invertase. These reactions also include two of the three essentially irreversible reactions of the glycolytic pathway that are catalyzed by hexokinase and phosphofructokinase. The phosphofructokinase reaction is one of the control points of glycolysis in both plants and animals.

Once glyceraldehyde- 3-phosphate is formed, the glycolytic pathway can begin to extract usable energy in the energy-conserving phase. The enzyme glyceraldehyde-3-phosphate dehydrogenase catalyzes the oxidation of the aldehyde to a carboxylic acid, reducing NAD^+ to NADH. This reaction releases sufficient free energy to allow the phosphorylation (using inorganic phosphate) of glyceraldehyde-3-phosphate to produce 1,3-bisphosphoglycerate. The phosphorylated carboxylic acid on carbon 1 of 1,3-bisphosphoglycerate has a large standard free energy of hydrolysis (-49.3 kJ mol^{-1}, or -11.8 kcal mol^{-1}). Thus, 1,3-bisphosphoglycerate is a strong donor of phosphate groups.

In the next step of glycolysis, catalyzed by phosphoglycerate kinase, the phosphate on carbon 1 is transferred to a molecule of ADP, yielding ATP and 3-phosphoglycerate. For each sucrose entering the pathway, four ATPs are generated by this reaction–one for each molecule of 1,3- bisphosphoglycerate. This type of ATP synthesis, traditionally referred to as substrate-level phosphorylation, involves the direct transfer of a phosphate group from a substrate molecule to ADP,

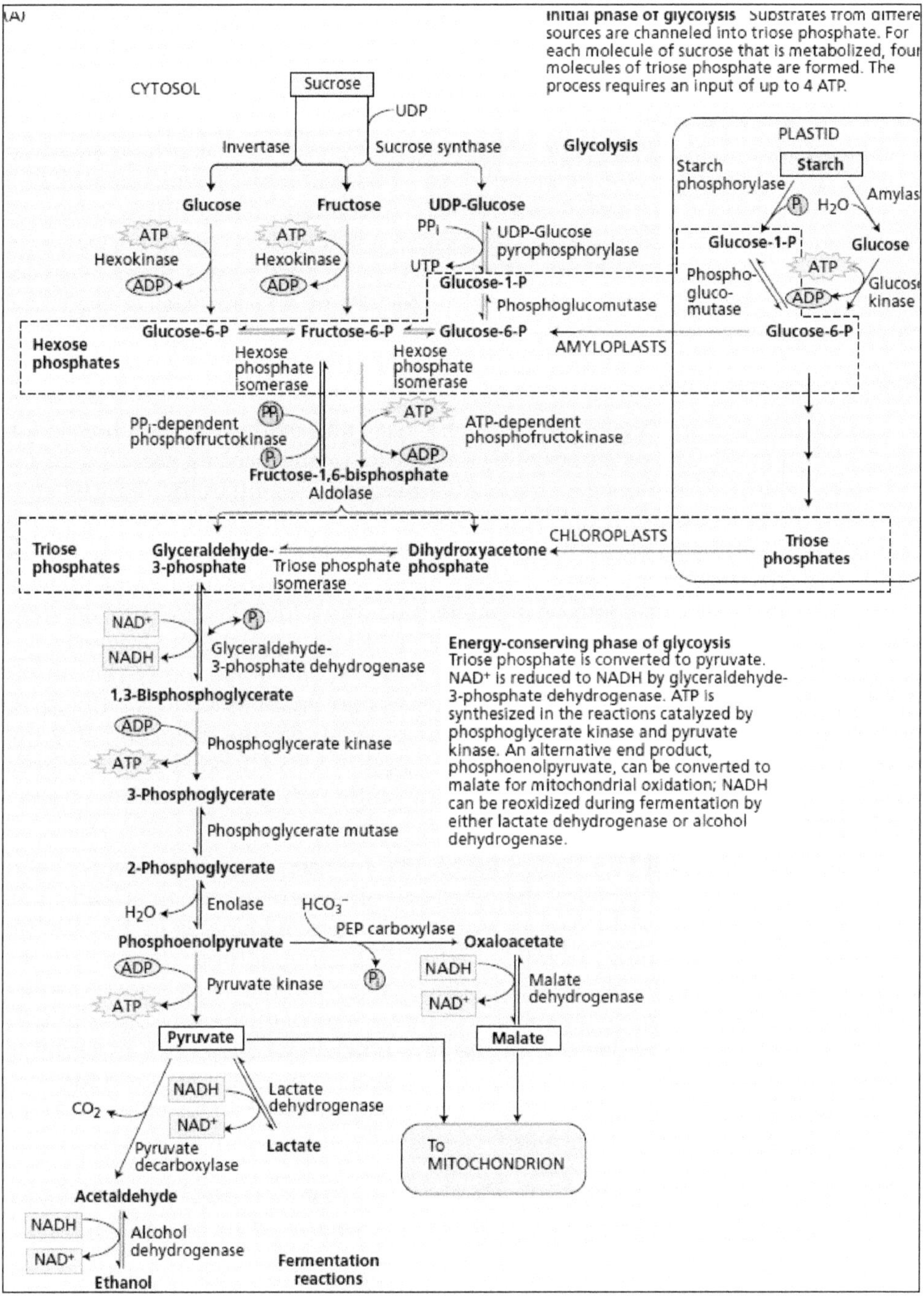

Figure 5.1: Glycolysis Cycle (Courtesy of Taiz and Zieger).

to form ATP. As we will see, ATP synthesis by substratelevel phosphorylation is mechanistically distinct from ATP synthesis by ATP synthases involved in the oxidative phosphorylation in mitochondria or photophosphorylation in chloroplasts.

In the following reaction, the phosphate on 3-phosphoglycerate is transferred to carbon 2 and a molecule of water is removed, yielding the compound phosphoenylpyruvate (PEP). The phosphate group on PEP has a high standard free energy of hydrolysis (–61.9 kJ mol–1, or –14.8 kcal mol–1), which makes PEP an extremely good phosphate donor for ATP formation. Using PEP as substrate, the enzyme pyruvate kinase catalyzes a second substrate-level phosphorylation to yield ATP and pyruvate. This final step, which is the third essentially irreversible step in glycolysis, yields four additional molecules of ATP for each sucrose that enters the pathway.

5.2. Fermentation

In the absence of oxygen, the citric acid cycle and oxidative phosphorylation cannot function. Glycolysis thus cannot continue to operate because the cell's supply of NAD+ is limited, and once all the NAD+ becomes tied up in the reduced state (NADH), the reaction catalyzed by glyceraldehyde- 3-phosphate dehydrogenase cannot take place. To overcome this problem, plants and other organisms can further metabolize pyruvate by carrying out one or more forms of fermentative metabolism. In alcoholic fermentation (common in plants, but more widely known from brewer's yeast), the two enzymes. pyruvate decarboxylase and alcohol dehydrogenase act on pyruvate, ultimately producing ethanol and CO_2 and oxidizing NADH in the process. Under some circumstances, plant tissues may be subjected to low (hypoxic) or zero (anoxic) concentrations of ambient oxygen, forcing them to carry out fermentative metabolism. The best-studied example involves flooded or waterlogged soils in which the diffusion of oxygen is sufficiently reduced to cause root tissues to become hypoxic.

The net synthesis of four molecules of ATP for each sucrose molecule that is converted to ethanol (or lactate), the efficiency of anaerobic fermentation is only about 4 per cent. Most of the energy available in sucrose remains in the reduced by-product of fermentation: lactate or ethanol. During aerobic respiration, the pyruvate produced by glycolysis is transported into mitochondria, where it is further oxidized, resulting in a much more efficient conversion of the free energy originally available in the sucrose. Because of the low efficiency of energy conservation under fermentation, an increased rate of glycolysis is needed to sustain the ATP production necessary for cell survival. This is called the ***Pasteur effect*** after the French microbiologist Louis Pasteur.

5.3. Pentose Phosphate Pathway (or oxidative pentose phosphate pathway or Hexose mono phosphate shunt)

The glycolytic pathway is not the only route available for the oxidation of sugars in plant cells. Sharing common metabolites, the oxidative pentose phosphate pathway (also known as the *hexose monophosphate shunt*) can also accomplish this task (Figure 5.2.4). The reactions are carried out by soluble enzymes present in the cytosol

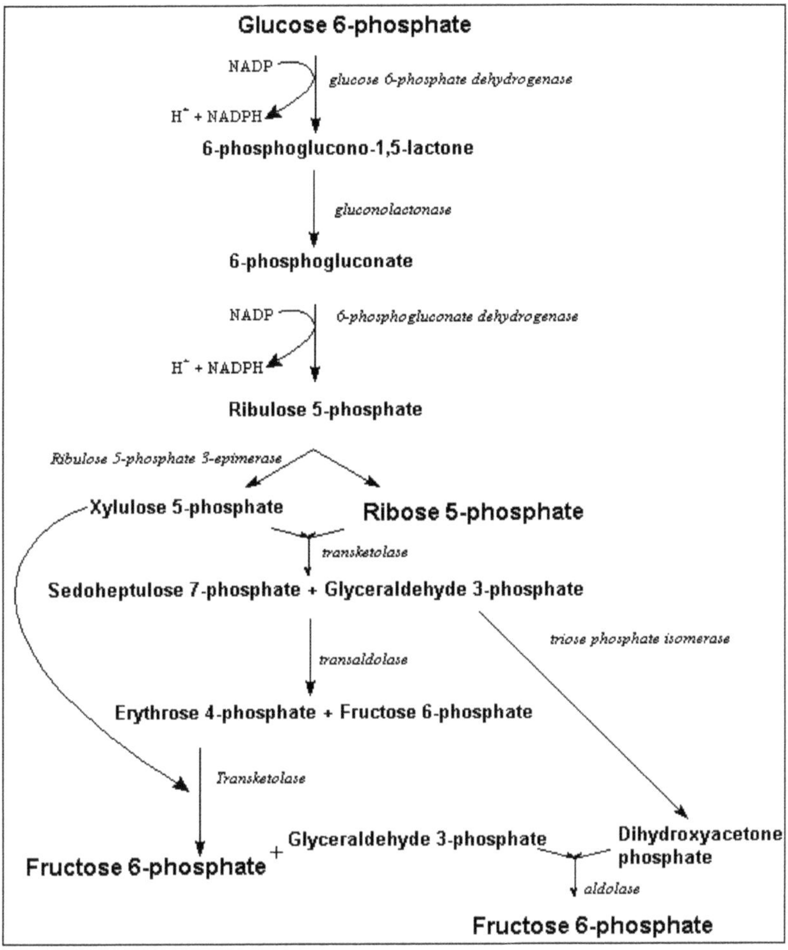

Figure 5.2: Pentose Phosphate Pathway.

and in plastids. Generally, the pathway in plastids predominates over the cytosolic pathway (Dennis *et al.*, 1997). The first two reactions of this pathway involve the oxidative events that convert the six-carbon glucose-6- phosphate to a five-carbon sugar, ribulose-5-phosphate, with loss of a CO_2 molecule and generation of two molecules of NADPH (not NADH). The remaining reactions of the pathway convert ribulose-5-phosphate to the glycolytic intermediates glyceraldehyde-3-phosphate and fructose-6-phosphate. Because glucose-6-phosphate can be regenerated from glyceraldehyde-3-phosphate and fructose-6- phosphate by glycolytic enzymes, for six turns of the cycle we can write the reaction as follows:

$$6 \text{ glucose-6-P} + 12 \text{ NADP}^+ + 7 H_2O \rightarrow \text{glucose-6-P} + 6 CO_2 + Pi + 12 \text{ NADPH} + 12 H^+$$

The net result is the complete oxidation of one glucose-6- phosphate molecule to CO_2 with the concomitant synthesis of 12 NADPH molecules. The oxidative pentose phosphate pathway plays several roles in plant metabolism:

- ☆ The product of the two oxidative steps is NADPH, and this NADPH is thought to drive reductive steps associated with various biosynthetic reactions that occur in the cytosol. In nongreen plastids, such as amyloplasts, and in chloroplasts functioning in the dark, the pathway may also supply NADPH for biosynthetic reactions such as lipid biosynthesis and nitrogen assimilation.
- ☆ Because plant mitochondria are able to oxidize cytosolic NADPH via an NADPH dehydrogenase localized on the external surface of the inner membrane, some of the reducing power generated by this pathway may contribute to cellular energy metabolism; that is, electrons from NADPH may end up reducing O_2 and generating ATP.
- ☆ The pathway produces ribose-5-phosphate, a precursor of the ribose and deoxyribose needed in the synthesis of RNA and DNA, respectively.
- ☆ Another intermediate in this pathway, the four-carbon erythrose-4-phosphate, combines with PEP in the initial reaction that produces plant phenolic compounds, including the aromatic amino acids and the precursors of lignin, flavonoids, and phytoalexins.
- ☆ During the early stages of greening, before leaf tissues become fully photoautotrophic, the oxidative pentose phosphate pathway is thought to be involved in generating Calvin cycle intermediates.

5.4. The Citric Acid Cycle

In 1937 the German-born British biochemist Hans A. Krebs reported the discovery of the citric acid cycle–also called the tricarboxylic acid cycle or Krebs cycle. The citric acid cycle is localized in the matrix of mitochondria, we will begin with a general description of mitochondrial structure and function.

The breakdown of sucrose to pyruvate releases less than 25 per cent of the total energy in sucrose; the remaining energy is stored in the two molecules of pyruvate. The next two stages of respiration (the citric acid cycle and oxidative phosphorylation–*i.e.*, electron transport coupled to ATP synthesis) take place within an organelle enclosed by a double membrane, the mitochondrion (plural *mitochondria*). Plant mitochondria–whether in situ or *in vitro*–usually look spherical or rod like. The ultrastructural features of plant mitochondria are similar to those of mitochondria in nonplant tissues (Figure 5.3). Plant mitochondria have two membranes: a smooth outer membrane that completely surrounds a highly invaginated inner membrane. The invaginations of the inner membrane are known as cristae (singular *crista*). As a consequence of the greatly enlarged surface area, the inner membrane can contain more than 50 per cent of the total mitochondrial protein. The aqueous phase contained within the inner membrane is referred to as the mitochondrial matrix (plural *matrices*), and the region between the two mitochondrial membranes is known as the intermembrane space. Intact mitochondria are osmotically active; that is, they take up water and swell when placed in a hypo-osmotic medium. Most inorganic ions and charged organic molecules are not able to diffuse freely into the matrix space. The inner membrane is the osmotic barrier; the outer membrane is permeable to solutes. Like chloroplasts, mitochondria are

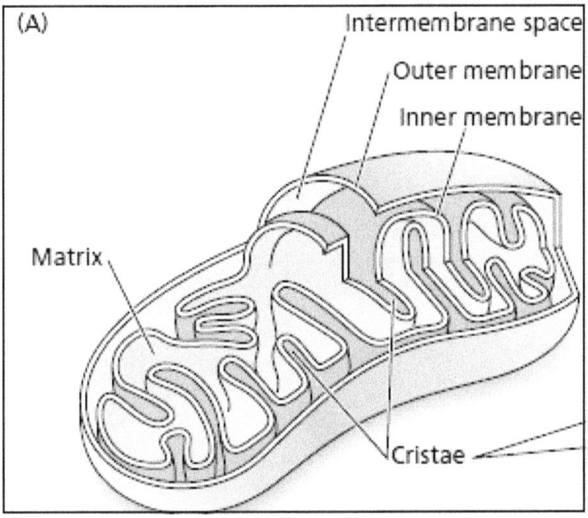

Figure: 5.3 Ultrastructure of Mitrochondria.

semiautonomous organelles because they contain ribosomes, RNA, and DNA, which encodes a limited number of mitochondrial proteins. Plant mitochondria are thus able to carry out the various steps of protein synthesis and to transmit their genetic information. Mitochondria proliferate through the division by fission of preexisting mitochondria and not through de novo biogenesis of the organelle.

The citric acid cycle is also known as the tricarboxylic acid cycle, because of the importance of the tricarboxylic acids citric acid (citrate) and isocitric acid (isocitrate) as early intermediates (Figure 5.4). This cycle constitutes the second stage in respiration and takes place in the mitochondrial matrix. Its operation requires that the pyruvate generated in the cytosol during glycolysis be transported through the impermeable inner mitochondrial membrane via a specific transport protein. Once inside the mitochondrial matrix, pyruvate is decarboxylated in an oxidation reaction by the enzyme pyruvate dehydrogenase. The products are NADH (from NAD^+), CO_2, and acetic acid in the form of acetyl-CoA, in which a thioester bond links the acetic acid to a sulfur-containing cofactor, coenzyme A (CoA).

Pyruvate dehydrogenase exists as a large complex of several enzymes that catalyze the overall reaction in a three-step process: decarboxylation, oxidation, and conjugation to CoA. In the next reaction the enzyme citrate synthase combines the acetyl group of acetyl-CoA with a four-carbon dicarboxylic acid (oxaloacetate, OAA) to give a six-carbon tricarboxylic acid (citrate). Citrate is then isomerized to isocitrate by the enzyme aconitase. The following two reactions are successive oxidative decarboxylations, each of which produces one NADH and releases one molecule of CO_2, yielding a four-carbon molecule, succinyl-CoA. At this point, three molecules of CO_2 have been produced for each pyruvate that entered the mitochondrion, or 12 CO_2 for each molecule of sucrose oxidized. During the remainder of the citric acid cycle, succinyl- CoA is oxidized to OAA, allowing the continued operation of the cycle. Initially the large amount of free energy available in the thioester bond

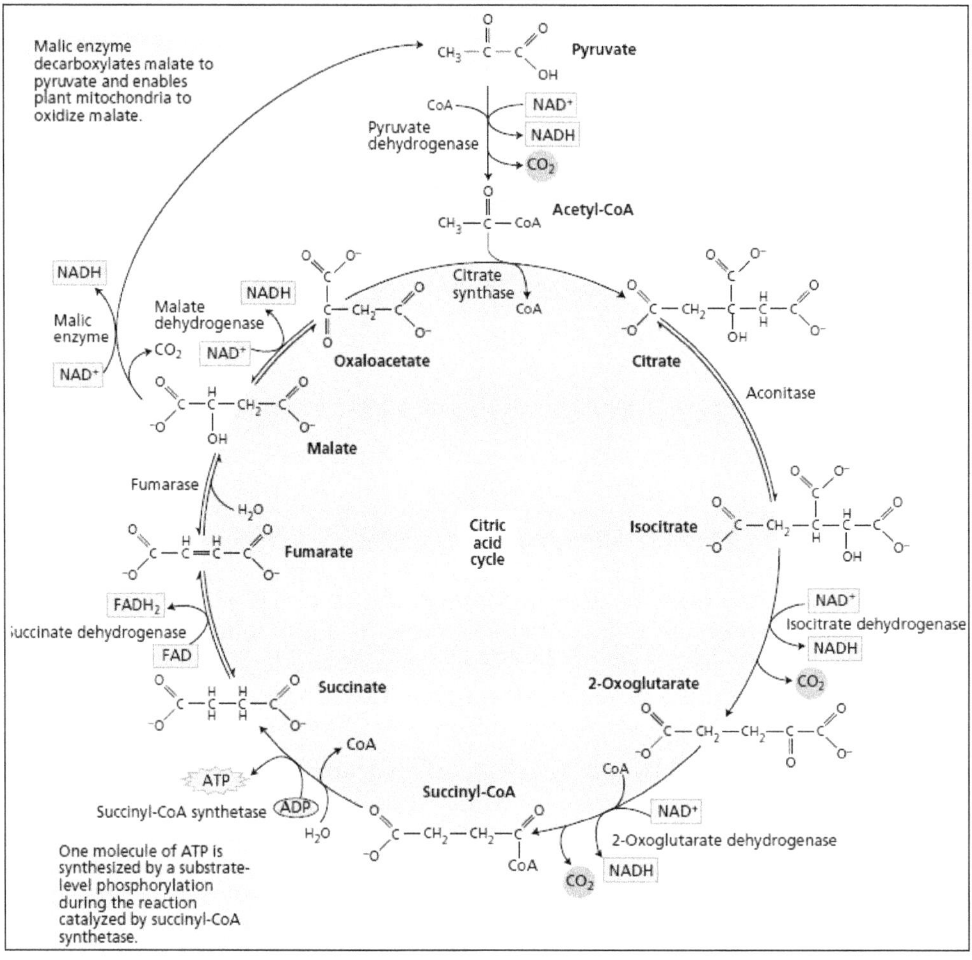

Figure 5.4: Citric Acid Cycle.

of succinyl-CoA is conserved through the synthesis of ATP from ADP and Pi via a substrate- level phosphorylation catalyzed by succinyl-CoA synthetase. (Recall that the free energy available in the thioester bond of acetyl-CoA was used to form a carbon– carbon bond in the step catalyzed by citrate synthase.) The resulting succinate is oxidized to fumarate by succinate dehydrogenase, which is the only membrane-associated enzyme of the citric acid cycle and also part of the electron transport chain (which is the next major topic to be discussed in this chapter). The electrons and protons removed from succinate end up not on NAD^+ but on another cofactor involved in redox reactions: FAD (flavin adenine dinucleotide). FAD is covalently bound to the active site of succinate dehydrogenase and undergoes a reversible two-electron reduction to produce $FADH_2$. In the final two reactions of the citric acid cycle, fumarate is hydrated to produce malate, which is subsequently oxidized by malate dehydrogenase to regenerate OAAand produce another molecule of NADH. The OAA produced is now able to react with another acetyl-CoA and continue the cycling.

The stepwise oxidation of one molecule of pyruvate in the mitochondrion gives rise to three molecules of CO_2, and much of the free energy released during these oxidations is conserved in the form of four NADH and one $FADH_2$. In addition, one molecule of ATP is produced by a substrate-level phosphorylation during the citric acid cycle. All the enzymes associated with the citric acid cycle are found in plant mitochondria. Some of them may be associated in multienzyme complexes, which would facilitate movement of metabolites between the enzymes.

5.5. Electron Transport and ATP Synthesis at the Inner Mitochondrial Membrane

ATP is the energy carrier used by cells to drive living processes, and chemical energy conserved during the citric acid cycle in the form of NADH and $FADH_2$ (redox equivalents with high-energy electrons) must be converted to ATP to perform useful work in the cell. This O_2-dependent process, called oxidative phosphorylation, occurs in the inner mitochondrial membrane. Although fundamentally similar in all aerobic cells, the electron transport chain of plants (and fungi) contains multiple NAD(P)H dehydrogenases and an alternative oxidase not found in mammalian mitochondria.

For each molecule of sucrose oxidized through glycolysis and the citric acid cycle pathways, 4 molecules of NADH are generated in the cytosol and 16 molecules of NADH plus 4 molecules of $FADH_2$ (associated with succinate dehydrogenase) are generated in the mitochondrial matrix. These reduced compounds must be reoxidized or the entire respiratory process will come to a halt. The electron transport chain catalyzes an electron flow from NADH (and $FADH_2$) to oxygen, the final electron acceptor of the respiratory process. For the oxidation of NADH, the overall two-electron transfer can be written as follows:

$$NADH + H^+ + 1D_2\,O_2 \longrightarrow NAD^+ + H_2O$$

From the reduction potentials for the NADH–NAD+ pair (–320 mV) and the H_2O–1D2O_2 pair (+810 mV), it can be calculated that the standard free energy released during this overall reaction ($-nF\,E\,2$) is about 220 kJ mol^{-1} (52 kcal mol^{-1}) per two electrons. The role of the electron transport chain is to bring about the oxidation of NADH (and $FADH_2$) and, in the process, utilize some of the free energy released to generate an electrochemical proton gradient across the inner mitochondrial membrane. The electron transport chain of plants contains the same set of electron carriers found in mitochondria from other organisms (Figure 5.5) (Siedow 1995; Siedow and Umbach 1995). The individual electron transport proteins are organized into four multiprotein complexes (identified by Roman numerals I through IV), all of which are localized in the inner mitochondrial membrane:

Complex I (NADH dehydrogenase)

Electrons from NADH generated in the mitochondrial matrix during the citric acid cycle are oxidized by complex I (an NADH dehydrogenase). The electron carriers in complex I include a tightly bound cofactor (flavin mononucleotide [FMN], which is chemically similar to FAD; see Figure 5.5) and several iron–sulfur centers. Complex I then transfers these electrons to ubiquinone. Four protons are

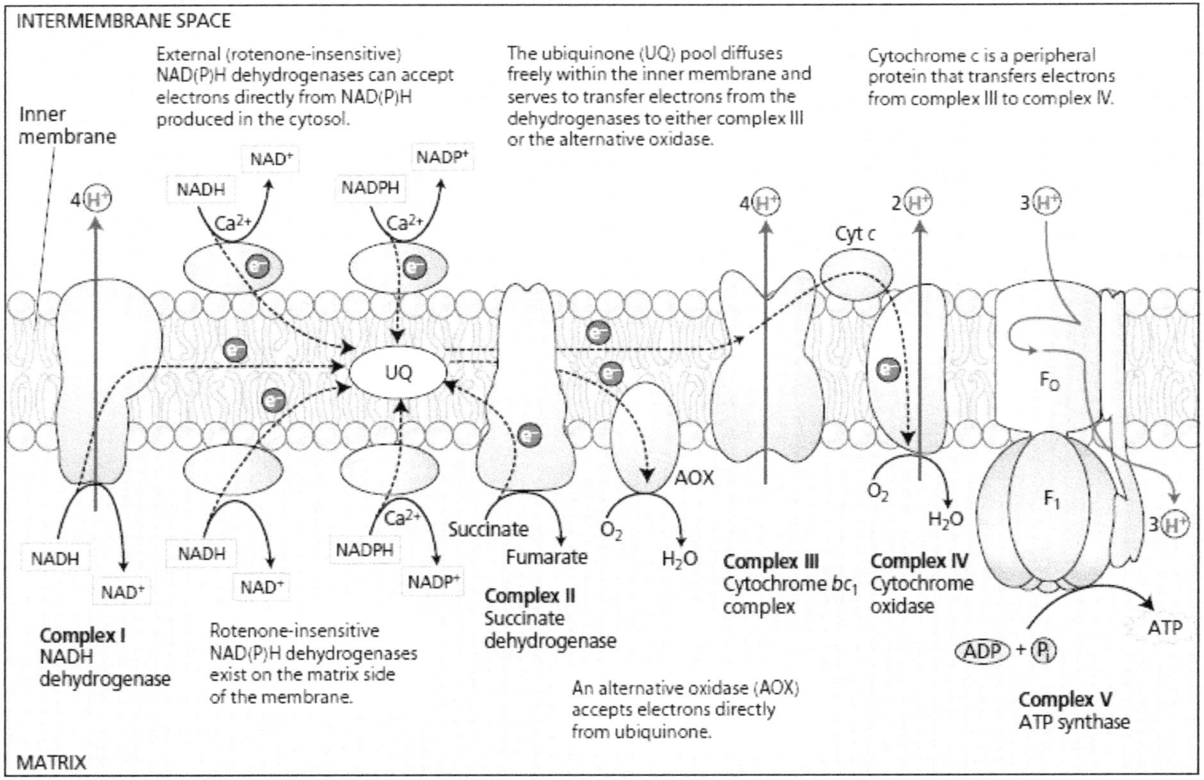

Figure 5.5: Organization of the Electron Transport Chain and ATP Synthesis.

Organization of the electron transport chain and ATP synthesis in the inner membrane of plant mitochondria. In addition to the five standard protein complexes found in nearly all other mitochondria, the electron transport chain of plant mitochondria contains five additional enzymes marked in green. None of these additional enzymes pumps protons. Specific inhibitors, rotenone for complex I, antimycin for complex III, cyanide for complex IV, and salicylhydroxamic acid (SHAM) for the alternative oxidase, are important tools to investigate the electron transport chain of plant mitochondria.(Courtesy of Taiz and Zieger).

pumped from the matrix to the intermembrane space for every electron pair passing through the complex.

Ubiquinone, a small lipid-soluble electron and proton carrier, is located within the inner membrane. It is not tightly associated with any protein, and it can diffuse within the hydrophobic core of the membrane bilayer.

Complex II (Succinate dehydrogenase)

Oxidation of succinate in the citric acid cycle is catalyzed by this complex, and the reducing equivalents are transferred via the $FADH_2$ and a group of iron–sulfur proteins into the ubiquinone pool. This complex does not pump protons.

Complex III (Cytochrome bc1 complex)

This complex oxidizes reduced ubiquinone (ubiquinol) and transfers the electrons via an iron–sulfur center, two *b*-type cytochromes (*b*565 and *b*560), and a membrane-bound cytochrome *c*1 to cytochrome *c*. Four protons per electron pair are pumped by complex III.

Cytochrome *c* is a small protein loosely attached to the outer surface of the inner membrane and serves as a mobile carrier to transfer electrons between complexes III and IV.

Complex IV (Cytochrome c oxidase)

This complex contains two copper centers (CuA and CuB) and cytochromes *a* and *a*3. Complex IV is the terminal oxidase and brings about the four-electron reduction of O_2 to two molecules of H_2O. Two protons are pumped per electon pair (Figure 5.5). Both structurally and functionally, ubiquinone and the cytochrome *bc*1 complex are very similar to plastoquinone and the cytochrome *b*6 *f* complex, respectively, in the photosynthetic electron transport chain.

In oxidative phosphorylation, the transfer of electrons to oxygen via complexes I to IV is coupled to the synthesis of ATP from ADP and Pi via the ATP synthase (complex V). The number of ATPs synthesized depends on the nature of the electron donor. In experiments conducted with the use of isolated mitochondria, electrons derived from internal (matrix) NADH give ADP:O ratios (the number of ATPs synthesized per two electrons transferred to oxygen) of 2.4 to 2.7. Succinate and externally added NADH each give values in the range of 1.6 to 1.8, while ascorbate, which serves as an artificial electron donor to cytochrome *c*, gives values of 0.8 to 0.9. Results such as these (for both plant and animal mitochondria) have led to the general concept that there are three sites of energy conservation along the electron transport chain, at I, III and IV. The number of H^+ pumped by complexes I, III, and IV and the cost of 4 H^+ for synthesizing one ATP. For instance, electrons from external NADH pass only complexes III and IV, so a total of 6 H^+ are pumped, giving 1.5 ATP (when the alternative oxidase pathway is not used). The mechanism of mitochondrial ATP synthesis is based on the chemiosmotic hypothesis, which was first proposed in 1961 by Nobel laureate Peter Mitchell as a general mechanism of energy conservation across biological membranes (Nicholls and Ferguson 2002). According to the chemiosmotic theory, the orientation of electron carriers within the mitochondrial inner membrane allows for the transfer of protons (H^+) across the inner membrane during electron flow. Numerous studies have confirmed that mitochondrial electron transport is associated with a net transfer of protons from the mitochondrial matrix to the intermembrane space (see Figure 5.5) (Whitehouse and Moore, 1995).

Because the inner mitochondria l membrane is impermeable to H^+, an electrochemical proton gradient can build up. The free energy associated with the formation of an electrochemical proton gradient (also referred to as a proton motive force, Δp, when expressed in units of volts) is made up of an electric transmembrane potential component (ΔE) and a chemical-potential component (Δ pH). ΔE results from the asymmetric distribution of a charged species (H^+) across the membrane, and pH is due to the proton concentration difference across the membrane. Because protons are translocated from the mitochondrial matrix to the intermembrane space, the resulting ΔE across the inner mitochondrial membrane is negative. This situation contrasts to that in the chloroplast, where almost all of the proton motive force across the thylakoid membrane is made up by a proton gradient. The free energy can be utilized to carry out chemical work (ATP synthesis). The Δp ?is coupled to the synthesis of ATP by an additional protein complex associated with the inner membrane, the FoF1-ATP synthase. The FoF1-ATP synthase (also called *complex V*) consists of two major components, F1 and Fo (Figure 5.5). F1 is a peripheral membrane protein complex that is composed of at least five different subunits and contains the catalytic site for converting ADP and Pi to ATP. This complex is attached to the matrix side of the inner membrane. Fo is an integral membrane protein complex that consists of at least three different polypeptides that form the channel through which protons cross the inner membrane. The passage of protons through the channel is coupled to the catalytic cycle of the F1 component of the ATP synthase, allowing the ongoing synthesis of ATP and the simultaneous utilization of the Δ p. For each ATP synthesized, 3 H^+ pass through the Fo from the intermembrane space to the matrix down the electrochemical proton gradient. A high-resolution X-ray structure pof most of the F1 complex of the mammalian mitochondrial ATP synthase supports a "rotational model" for the catalytic mechanism of ATP synthesis.

The structure and function of the mitochondrial ATP synthase is similar to that of the CFo–CF1 ATP synthase in photophosphorylation. The operation of a chemiosmotic mechanism of ATP synthesis has several implications. First, the true site of ATP formation on the mitochondrial inner membrane is the ATP synthase, not complex I, III, or IV. These complexes serve as sites of energy conservation whereby electron transport is coupled to the generation of a Δp Second, the chemiosmotic theory explains the action mechanism of uncouplers, a wide range of chemically unrelated compounds (including 2,4-dinitrophenol and FCCP[*p* trifluoromethoxycarbonylcyanide phenylhydrazone]) that decreases mitochondrial ATP synthesis but often stimulates the rate of electron transport. All of these compounds make the inner membrane leaky to protons, which prevents the build up of a sufficiently large Δpto drive ATP synthesis. (Figure 5.5). ADP provides a substrate that stimulates dissipation of the Δp the through the FoF1-ATP synthase during ATP synthesis. Once all the ADP has been converted to ATP, the Δp builds up again and reduces the rate of electron flow (*state* 4). The ratio of the rates with and without ADP (state 3:state 4) is referred to as the *respiratory control ratio.*

The electrochemical proton gradient also plays a role in the movement of the organic acids of the citric acid cycle and of substrates and products of ATP synthesis in and out of mitochondria. Although ATP is synthesized in the mitochondrial matrix, most of it is used outside the mitochondrion, so an efficient mechanism is needed for moving ADP in and ATP out of the organelle. The uptake of inorganic phosphate (Pi) involves an active phosphate transporter protein that uses the proton gradient component (pH) of the proton motive force to drive the electroneutral exchange of Pi – (in) for OH^- (out). As long as a pH is maintained across the inner membrane, the Pi content within the matrix will remain high. Similar reasoning applies to the uptake of pyruvate, which is driven by the electroneutral exchange of pyruvate for OH–, leading to continued uptake of pyruvate from the cytosol (Figure 5.6). The total cost of taking up a phosphate (1 OH^- out, which is the same as 1 H^+ in) and exchanging ADP for ATP (one negative charge out, which is the same as one positive charge in) is 1 H^-. This proton should also be included in calculation of the cost of synthesizing one ATP. Thus the total cost is 3 H^+ used by the ATP synthase plus 1 H^+ for the exchange across the membrane, or a total of 4 H^+.

5.6. Balance Sheet of ATP

ATP is regarded as " energy rich" or " high energy" compounds because it shows a large decrease in free energy when undergoes hydrolysis. These energy can be utilized by many metabolic activities.

Stage	*Release of ATP per Hexose Mol.*	*ATP Consumed*	*Net Gain of ATP*
Glycolysis			
1. Glucose → G-6-p		1	
2. Fructose-6-P → Fructose1,6-P		1	
3. 1,3-bisphosphoglyceraldehyde → 1,3-bisphosphoglyceric acid	2x2 = 4		
4. 1,3-bisphosphoglyceric acid → 3-phosphoglyceric acid	1x2 = 2		
5. 2-phosphoenol pyruvic acid → Pyruvic acid	1x2 = 2		
Total	**8**	**–2**	**= 6**
Krebs Cycle			
6. Pyruvic acid → Acetyl Co-A	3x2 = 6		
7. Iso-citric acid → Oxalosuccinic acid	3x2 = 6		
8. α-ketoglutaric acid → Succinyl Co-A	3x2 = 6		
9. Succinyl Co~A → Succinic acid	1x2 = 2		
10. Succinic acid → Fumaric acid	2x2 = 4		
11. Malic acid → Oxaloacetic acid	3x2 = 6		
Total	**30**	**0**	
Grand Total	**38**	**–2**	**= 36**

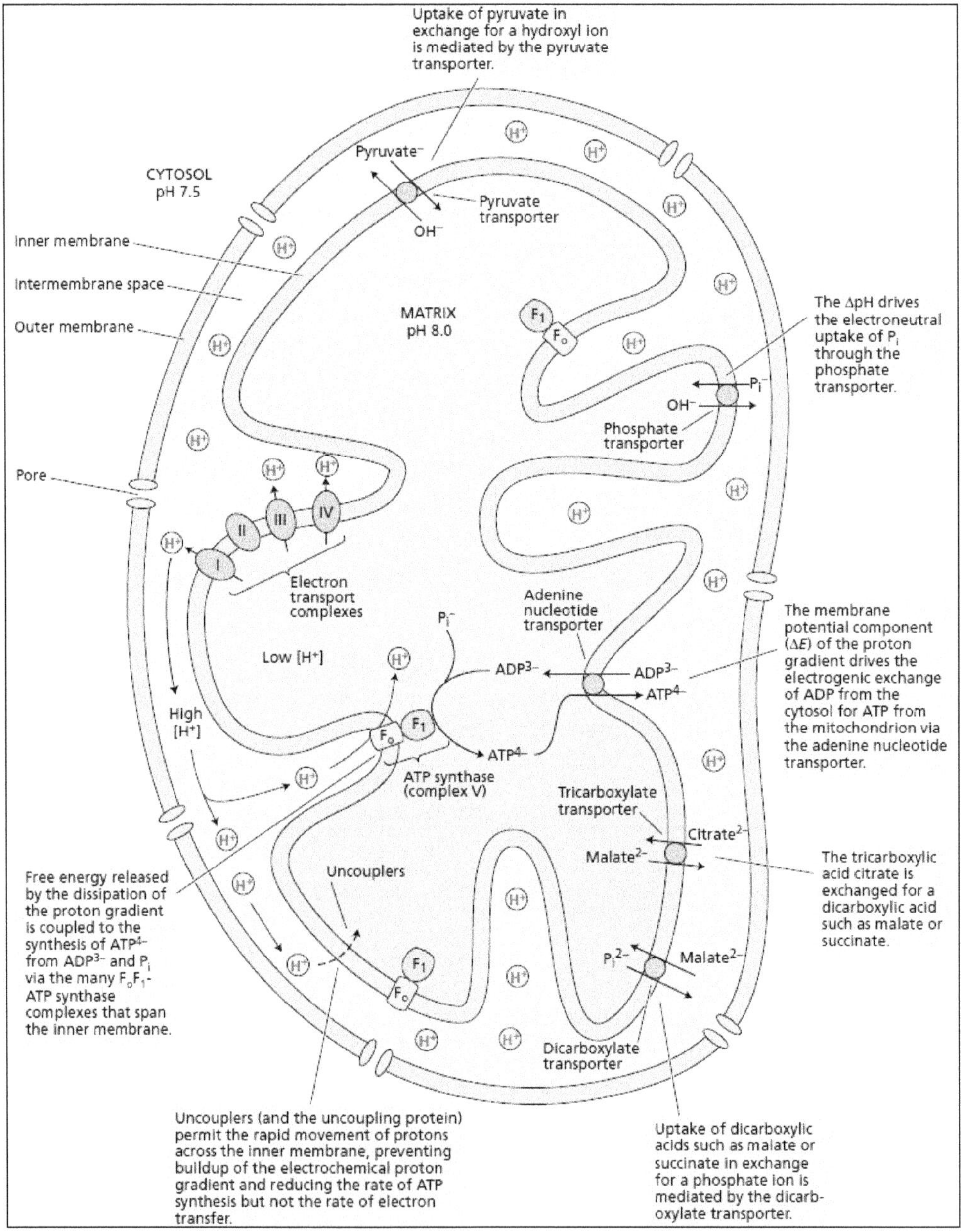

Figure 5.6: Transmembrane Transport in Plant Mitochondria. Electrochemical proton gradient established across inner mitochondrial membrane. (Courtesy of Taiz and Zieger).

Total gain of ATP molecules produced by the oxidative breakdown of glucose = 36

Alternative Respiration Systems (Role of the nonphosphorylating mechanisms)

The Alternative Oxidase

If cyanide is added to actively respiring animal tissues, cytochrome *c* oxidase is inhibited and the respiration rate quickly drops to less than 1 per cent of its initial level. However, most plant tissues display a level of cyanide-resistant respiration that can represent 10 to 25 per cent, and in some tissues up to 100 per cent, of the uninhibited control rate. The enzyme responsible for this oxygen uptake has been identified as a cyanide-resistant oxidase component of the plant mitochondrial electron transport chain called the alternative oxidase. Electrons feed off the main electron transport chain into the alternative pathway at the level of the ubiquinone pool. The alternative oxidase, the only component of the alternative pathway, catalyzes a four-electron reduction of oxygen to water and is specifically inhibited by several compounds, most notably salicylhydroxamic acid (SHAM). When electrons pass to the alternative pathway from the ubiquinone pool, two sites of proton pumping (at complexes III and IV) are by passed. Because there is no energy conservation site in the alternative pathway between ubiquinone and oxygen, the free energy that would normally be conserved as ATP is lost as heat when electrons are shunted through the alternative pathway. How can a process as seemingly energetically wasteful as the alternative pathway contribute to plant metabolism? One example of the functional usefulness of the alternative oxidase is its activity during floral development in certain members of the Araceae (the arum family)–for example, the voodoo lily (*Sauromatum guttatum*). Just before pollination, tissues of the clublike inflorescence, called the *appendix*, which bears male and female flowers, exhibit a dramatic increase in the rate of respiration via the alternative pathway. As a result, the temperature of the upper appendix increases by as much as 25°C over the ambient temperature for a period of about 7 hours. During this extraordinary burst of heat production, certain amines, indoles, and terpenes are volatilized, and the plant therefore gives off a putrid odor that attracts insect pollinators. Salicylic acid, a phenolic compound related to aspirin has been identified as the chemical signal responsible for initiating this thermogenic event in the voodoo lily. In most plants, however, both the respiratory rates and the rate of cyanide-resistant respiration are too low to generate sufficient heat to raise the temperature significantly, so what other role(s) does the alternative pathway play? It has been suggested that the alternative pathway can function as an "energy overflow" pathway, oxidizing respiratory substrates that accumulate in excess of those needed for growth, storage, or ATP synthesis (Lambers, 1985). This view suggests that electrons flow through the alternative pathway only when the activity of the main pathway is saturated. Such saturation is reached in the test tube in state 4; in vivo, saturation may occur if the respiration rate exceeds the cell's demand for ATP (*i.e.*, if ADP levels are very low). However, it is now clear that the alternative oxidase can be active before the cytochrome pathway is saturated. Thus the alternative oxidase makes it possible for the mitochondrion to adjust the relative rates of ATP production and synthesis of carbon skeletons for use in biosynthetic reactions. Another possible function of the alternative pathway is in the response of plants to a variety of stresses (phosphate deficiency, chilling, drought, osmotic stress, and so on), many of which can inhibit mitochondrial respiration (Wagner and Krab 1995). By draining off electrons from the electron transport chain, the alternative pathway prevents a potential overreduction of the ubiquinone pool, which, if left unchecked, can lead to the generation of destructive reactive oxygen species such as superoxide anions and hydroxyl radicals. In this way the alternative pathway may lessen the detrimental effects of stress on respiration (Wagner and Krab 1995; Møller 2001).

5.7. The Uncoupling Protein

A protein found in the inner membrane of mammalian mitochondria, the uncoupling protein, can dramatically increase the proton permeability of the membrane and thus act as an uncoupler. As a result, less ATP and more heat is generated. Heat production appears to be one of the uncoupling protein's main functions in mammalian cells. It has long been thought that the alternative oxidase in plants and the uncoupling protein in mammals were simply two different means of achieving the same end. It was therefore surprising when a protein similar to the uncoupling protein was discovered in plant mitochondria (Vercesi *et al.*, 1995; Laloi *et al.*, 1997). This protein is stress induced and, like the alternative oxidase, may function to prevent over reduction of the electron transport chain. It remains unclear, however, why plant mitochondria require both mechanisms.

The Internal, Rotenone-insensitive NADH Dehydrogenase, NDin (NADH)

This is one of the multiple NAD(P)H dehydrogenases found in plant mitochondria.

It has been suggested to work as a nonproton- pumping bypass when complex I is overloaded. Complex I has a higher affinity for NADH (ten times lower *Km*), than NDin(NADH). At lower NADH levels in the matrix, typically when ADP is available (state 3), complex I will dominate, whereas when ADP is rate limiting (state 4), NADH levels will increase and NDin(NADH) will be more active. The physiological importance of this enzyme is, however, still unclear.

5.8. Storage Lipids are Converted into Carbohydrates in Germinating Seeds (Glyoxylate Cycle)

After germinating, oil-containing seeds metabolize stored triacylglycerols by converting lipids to sucrose. Plants are not able to transport fats from the endosperm to the root and shoot tissues of the germinating seedling, so they must convert stored lipids to a more mobile form of carbon, generally sucrose. This process involves several steps that are located in different cellular compartments: oleosomes, glyoxysomes, mitochondria, and cytosol.

Overview: Lipids to Sucrose

The conversion of lipids to sucrose in oilseeds is triggered by germination and begins with the hydrolysis of triacylglycerols stored in the oil bodies to free fatty acids, followed by oxidation of the fatty acids to produce acetyl-CoA (Figure 5.7). The fatty acids are oxidized in a type of peroxisome called a glyoxysome, an

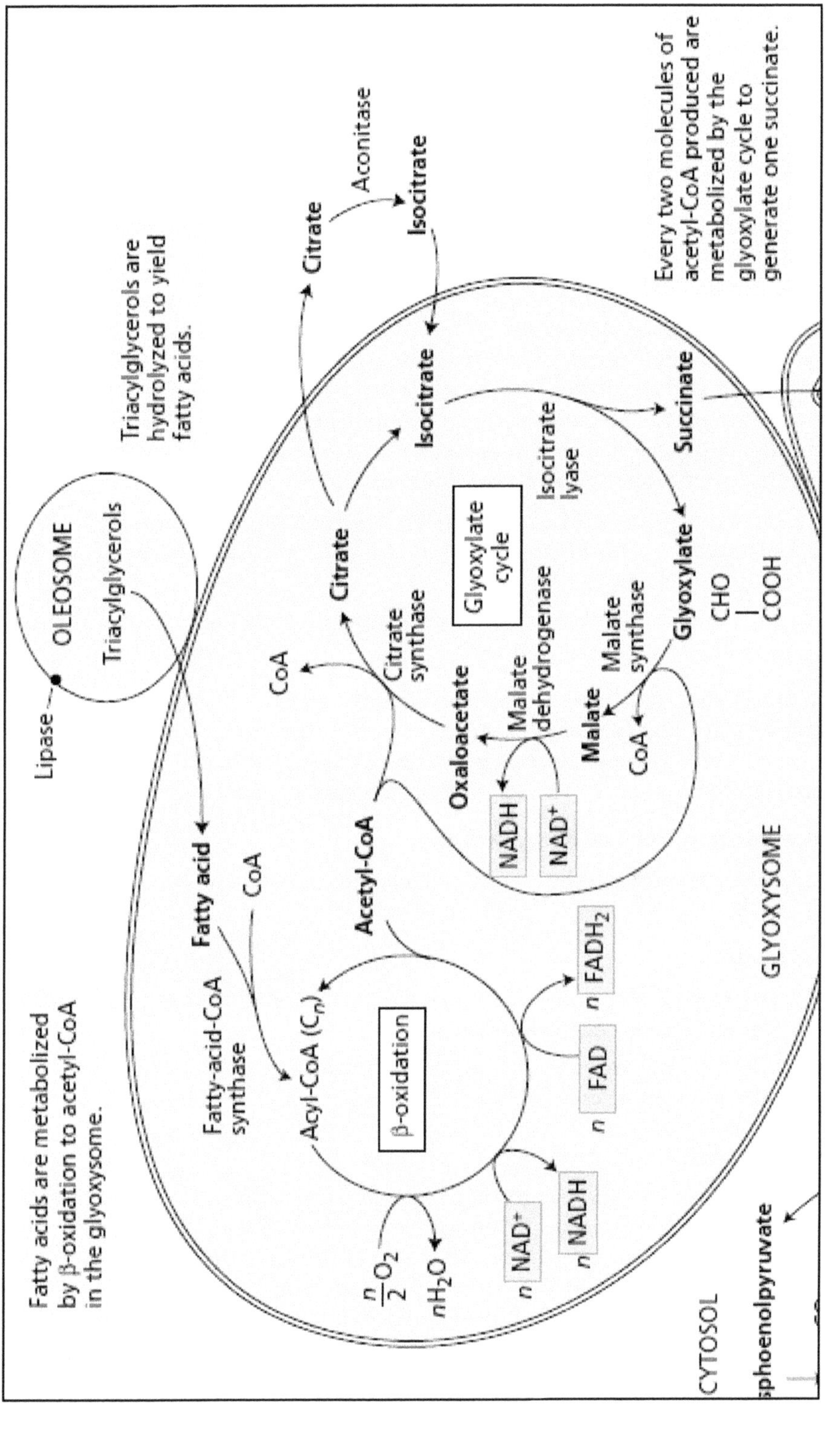

Figure 5.7: Malate Aspartate Cycle.

organelle enclosed by a single bilayer membrane that is found in the oil-rich storage tissues of seeds. Acetyl-CoA is metabolized in the glyoxysome (see Figure 5.7) to produce succinate, which is transported from the glyoxysome to the mitochondrion, where it is converted first to oxaloacetate and then to malate. The process ends in the cytosol with the conversion of malate to glucose via gluconeogenesis, and then to sucrose. Although some of this fatty acid–derived carbon is diverted to other metabolic reactions in certain oilseeds, in castor bean (*Ricinus communis*) the process is so efficient that each gram of lipid metabolized results in the formation of 1 g of carbohydrate.

The function of the glyoxylate cycle is to convert two molecules of acetyl-CoA to succinate.The acetyl-CoA produced by ?-oxidation is further metabolized in the glyoxysome through a series of reactions that make up the glyoxylate cycle (see Figure 5.7). Initially, the acetyl-CoA reacts with oxaloacetate to give citrate, which is then transferred to the cytoplasm for isomerisation to isocitrate by aconitase. Isocitrate is re imported into the peroxisome and converted to malate by two reactions that are unique to the glyoxylate pathway.

1. First isocitrate (C6) is cleaved by the enzyme isocitrate lyase to give succinate (C4) and glyoxylate (C2). This succinate is exported to the motochondria.
2. Next malate synthase combines a second molecule of acetyl-CoA with glyoxylate to produce malate.

Malate is then oxidized by malate dehydrogenase to oxaloacetate, which can combine with another acetyl-CoA to continue the cycle (see Figure 5.7). The glyoxylate produced keeps the cycle operating in the glyoxysome, but the succinate is exported to the mitochondria for further processing.

Factors Affecting Rate of Respiration

Many external and internal factors effects the rate of respiration. These factors are as follows:

Internal Factors

1. The rate of respiration depends on the amount, concentration and the state of activity of protoplasm.
2. When ample amount of respiratory substrates are present in the cell,the rate of respiration is high.

External Factors

- ✰ Temperature - Most of the plants respire normally between 5°C and 25°C. With further increases in temperature up to 30°C to 35°C the respiration rate increases. Further increases in temperature causes decreses in respiration rate. Still higher temperature may cause denaturation of respiratory enzyme and ultimately inhibit respiration.
- ✰ Light - Effects of light on the rate of respiration is indirects.
- ✰ Available oxygen - It is important for aerobic respiration.

- ✰ Water - If properly watered, the rate of respiration optimum. The rate become low when water available low, because the respiratory enzyme become inactive in the abscence of of this medium.
- ✰ Inorganic salts- in presence of salts, rate increases.
- ✰ Injury – It causes increases in the rate of respiration.

References

Buchanan, B.B., Gruissem, W., and Jones, R.L. (2000). *Biochemistry and Molecular Biology of Plants.* Rockville, MD: American Society of Plant Physiologists.

Foyer, C.H., Ferrario-Mery, S., and Huber, S.C. (2000). Regulation of carbon fluxes in the cytosol: Coordination of sucrose synthesis, nitrate reduction and organic and amino acid biosynthesis. In: *Photosynthesis: Physiology and Metabolism* R,C, Leegood, T.D. Sharkey, S. von Caemerer, (eds), *Advances in Photosynthesis and Respiration*, Vol. 9, pp. 177-203. Dordrecht: Kluwer.

Vanlerbrghe, G.C. and Ordog, S.H. (2002). Alternative oxidase: Integrating carbon metabolism and electron transport in plant respiration. *Photosynthesis: Physiology and Metabolism.* In: R.C. Leegood, T.D. Sharkey, S. von Caemmerer (eds.), *Advances in Photosynthesis and Resiration*,Vol. 9, pp. 173-191. Dordrecht: Kluwer.

Chapter 6

Plant Hormone

Plants also produce signaling molecules as organic substances, called *hormones*, that have profound effects on development at vanishingly low concentrations. Until quite recently, plant development was thought to be regulated by only five types of hormones: auxins, gibberellins, cytokinins, ethylene, and abscisic acid. However, there is now compelling evidence for the existence of plant steroid hormones, the brassinosteroids, that have a wide range of morphological effects on plant development.

6.1 Auxin

The auxin is indole-3- acetic acid (IAA). Several other auxins in higher plants were discovered later, but IAA is by far the most abundant and physiologically relevant. Because the structure of IAA is relatively simple.

Types of Auxin

1. Natural Auxin

Example- IAA, 4-chloro IAA etc. (Figure 6.1).

2. Synthetic Auxin

They are not treated as phytohormones, but they considered 'plant growth regulators'. Example - IPA, NAA, 2,4-D, 2,4,5-T, MCPA, 2-methoxy-3,6-dichlorobenzoic acid (Dicamba) etc.

A bioassay is a measurement of the effect of a known or suspected biologically active substance on living material. Went used *Avena sativa* (oat) coleoptiles in a technique called the ***Avena*** coleoptile curvature test.The coleoptiles curved because the increase in auxin on one side stimulated cell elongation, and the decrease in auxin on the other side (due to the absence of the coleoptile tip) caused a decrease in the growth rate–a phenomenon called differential growth. Went found that he could estimate the amount of auxin in a sample by measuring the resulting

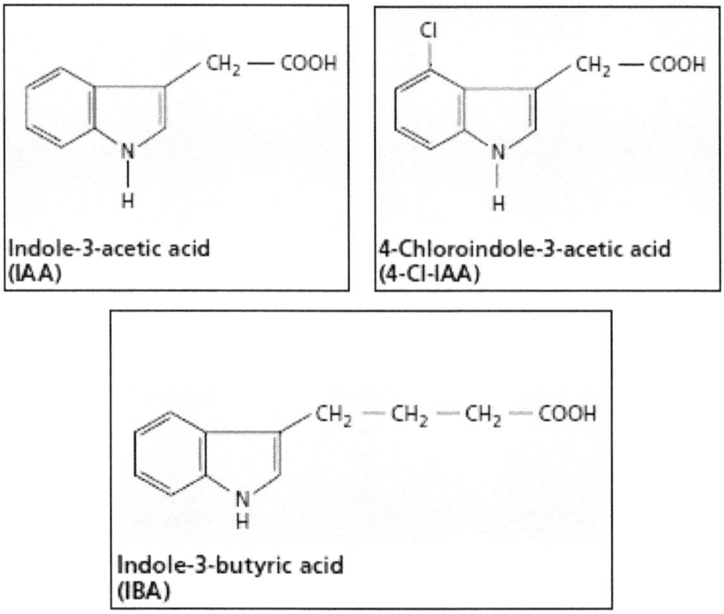

Figure 6.1: Structure of Three Natural Auxin.

coleoptile curvature. Auxin bioassays are still used today to detect the presence of auxin activity in a sample.

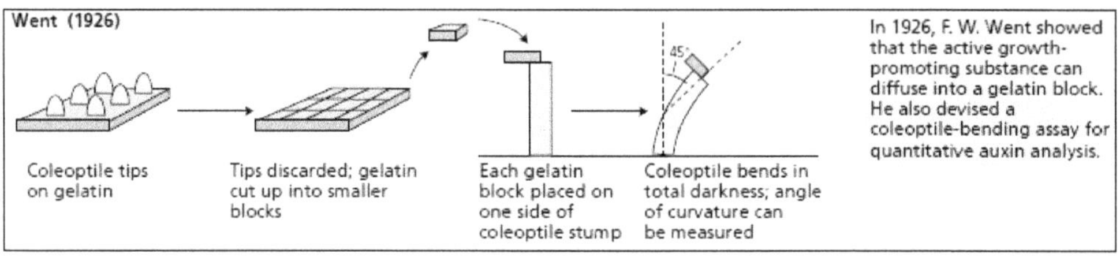

Figure 6.2: Diagrammatic Summary of Went's Experiment in the Discovery of Auxin.

6.1.1. Biosynthesis

IAA biosynthesis is associated with rapidly dividing and rapidly growing tissues, especially in shoots apices. Although virtually all plant tissues appear to be capable of producing low levels of IAA, shoot apical meristems, young leaves, and developing fruits and seeds are the primary sites of IAA synthesis. Auxin biosynthesis focused on amino acid tryptophan as the precursor.

1. First of all tryptophan undergo transamination reaction.
2. Then undergoes decarboxylation to become indole acetaldehyde.
3. Finally it oxidise to become IAA.

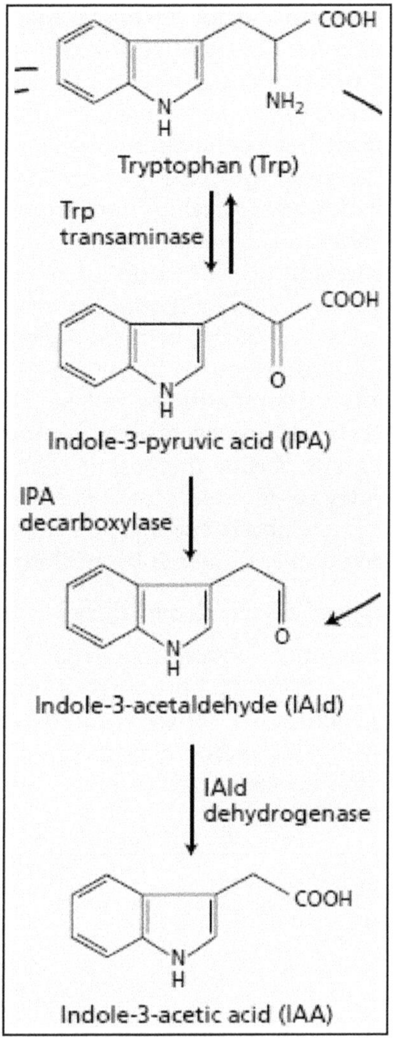

Figure 6.3: Biosynthesis of IAA.

IAA Breakdown

The enzyme system involved in IAA destruction is called IAA oxidase. Thus it is in multiple enzyme form. The activity of enzyme consume one molecule of o and release of one molecule of co for every molecule of IAA destructed. IAA.Oxidase activity is generally higher in the older, nongrowing tissue than it is in younger, actively growing tissues, which have high auxin requirement.

6.1.2. Auxin Transport

The main axis of shoots and roots, along with their branches, exhibit apex–base structural polarity, and this structural polarity has its origin in the polarity of auxin transport. Soon after Went developed the coleoptile curvature test for auxin, it was discovered that IAA moves mainly from the apical to the basal end (basipetally) in excised oat coleoptile sections. This type of uni -directional transport is termed polar transport. Auxin is the only plant growth hormone known to be transported polarly. Because the shoot apex serves as the primary source of auxin for the entire plant, polar transport has long been believed to be the principal cause of an auxin gradient extending from the shoot tip to the root tip. The longitudinal gradient of auxin from the shoot to the root affects various developmental processes, including stem elongation, apical dominance, wound healing, and leaf senescence. Recently it has been recognized that a significant amount of auxin transport also occurs in the phloem, and that the phloem is probably the principal route by which auxin is transported acropetally (*i.e.*, toward the tip) in the root. Thus, more than one pathway is responsible for the distribution of auxin in the plant. Polar transport proceeds in a cell-to-cell fashion, rather than via the symplast. That is, auxin exits the cell through the plasma membrane, diffuses across the compound middle lamella, and enters the cell below through its plasma membrane. The loss of auxin from cells is termed auxin efflux; the entry of auxin into cells is called auxin uptake or influx. The overall process requires metabolic energy, as evidenced by the sensitivity of polar transport to O_2 deprivation and metabolic inhibitors.

Inhibitors of Auxin Transport Block Auxin Efflux

Several compounds have been synthesized that can act as auxin transport inhibitors (ATIs), including NPA (1-Nnaphthylphthalamic acid) and TIBA (2,3,5-triiodobenzoic acid) (Figure 6.4). These inhibitors block polar transport by preventing auxin efflux. We can demonstrate this phenomenon by incorporating NPA or TIBA into either the donor or the receiver block in an auxin transport

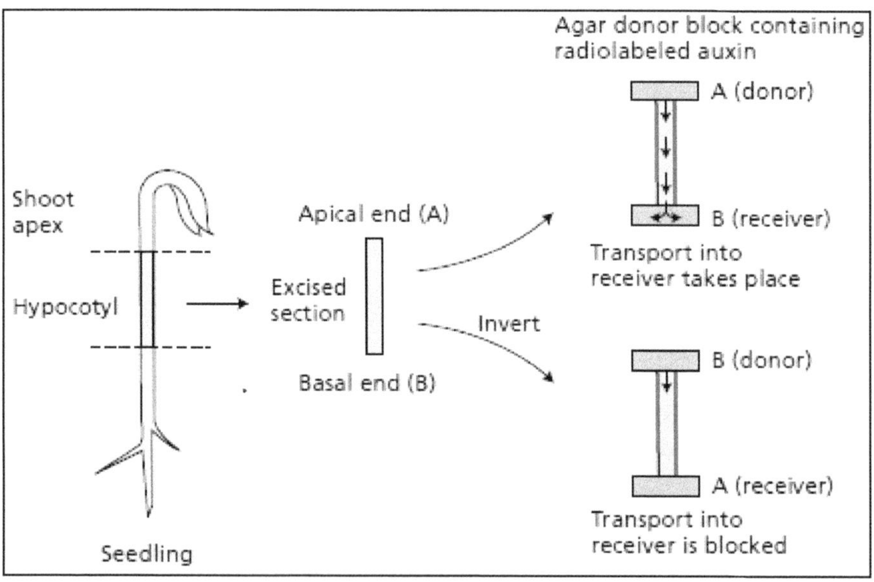

Figure 6.4: Method for Measuring Polar Auxin Transport. The polarity of transport is independent of orientation with respect to gravity.

experiment. Both compounds inhibit auxin efflux into the receiver block, but they do not affect auxin uptake from the donor block.

6.1.3. Physiological Effects of Auxin

1. Cell elongation and longitudinal growth
2. Cell divison in the cambium
3. Cell divison in tissue culture
4. Root growth and root initiation

Auxin promote root growth only at low concentration (*i.e.* 10 to 10 M concentration).At higher concentration they always inhibit root growth. In fact the concentration of auxin, which is most suitable for shoot elongation, is always inhibitory to root growth. High auxin concentration will promote initiation of secondary or branch roots. Secondary root iniation is normally controlled by auxin supplied by the shoot. Auxin also promote adventitious root formation on stems

6.1.4. Apical Dominance

The influence of apical bud in suppressing the growth of lateral buds is termed as apical dominance.auxin synthesized in apical meristem.

Prevention of Abscission Layer

A high concentration of auxin prevents the formation of abscission layer and the phenomenon is controlled by the concentration of aauxin and ethylene.

Auxin induces parthenocarpy in small number of plants, particularly in the families solanaceae (tomato, pepper), Cucurbitae and in citrus.

Auxin generally inhibit the flowing but in pineapple initiates flowering.

Auxin Used as Herbicides

Many synthetic auxins such as 2,4-D etc. are used as herbicides.

6.2. Gibberellins

Rice farmers in Asia had long known of a disease that makes the rice plants grow tall but eliminates seed production. In Japan this disease was called the "foolish seedling," or *bakanae*, disease. Plant pathologists investigating the disease found that the tallness of these plants was induced by a chemical secreted by a fungus that had infected the tall plants. This chemical was isolated from filtrates of the cultured fungus and called *gibberellin* after *Gibberella fujikuroi*, the name of the fungus. In the 1930s Japanese scientists succeeded in obtaining impure crystals of two fungal growth-active compounds, which they termed *gibberellin A* and *B*.

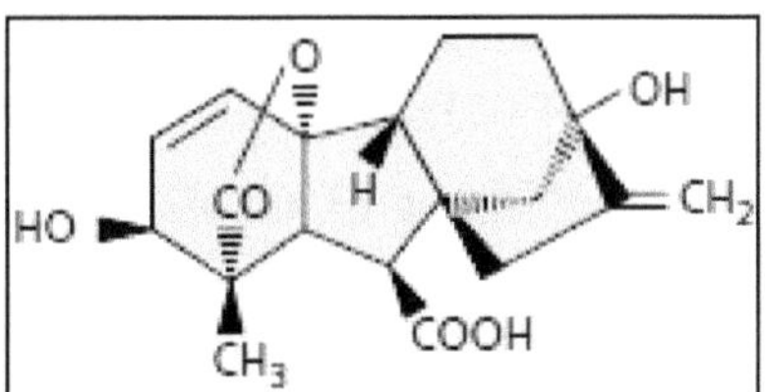

Gibberellic acid (GA_3)

Figure 6.6

6.2.1. Biosynthesis and Metabolism

Gibberellins constitute a large family of diterpene acids and are synthesized by a branch of the terpenoid pathway (mevalonic acid pathway). Gibberellins are tetracyclic diterpenoids made up of four isoprenoid units. Terpenoids are compounds made up of five-carbon (isoprene) building blocks:

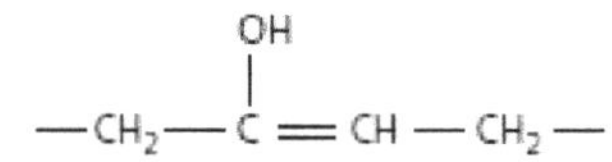

The gibberellin biosynthetic pathway can be divided into three stages, each residing in a different cellular compartment.

Stage 1: Production of Terpenoid Precursors and Ent-kaurene in Plastids

The basic biological isoprene unit is isopentenyl diphosphate (IPP). IPP used in gibberellin biosynthesis in green tissues is synthesized in plastids from glyceraldehyde-3-phosphate and pyruvate. Once synthesized, the IPPisoprene units are added successively to produce intermediates of 10 carbons (geranyl diphosphate), 15 carbons (farnesyl diphosphate), and 20 carbons (geranylgeranyl diphosphate, GGPP). GGPP is a precursor of many terpenoid compounds, including carotenoids and many essential oils, and it is only after GGPP that the pathway becomes specific for gibberellins. The cyclization reactions that convert GGPP to ent-kaurene represent the first step that is specific for the gibberellins.

Stage 2: Oxidation Reactions on the ER form GA12 and GA53

In the second stage of gibberellin biosynthesis, a methyl group on *ent*-kaurene is oxidized to a carboxylic acid, followed by contraction of the B ring from a six- to a five-carbon ring to give GA12-aldehyde. GA12-aldehyde is then oxidized to GA12, the first gibberellin in the pathway in all plants and thus the precursor of all the other gibberellins.

Stage 3: Formation in the Cytosol of all other Gibberellins from GA12 or GA53

All subsequent steps in the pathway (Figure 6.7) are carried out by a group of soluble dioxygenases in the cytosol. These enzymes require 2- oxoglutarate and molecular oxygen as cosubstrates, and they use Fe^{2+} and ascorbate as cofactors.

6.2.3. Effects of Gibberellin

Gibberellins Stimulate Stem Growth in Dwarf and Rosette Plants

Applied gibberellin promotes internodal elongation in a wide range of species. However, the most dramatic stimulations are seen in dwarf and rosette species, as well as members of the grass family. Exogenous GA_3 causes such extreme stem elongation in dwarf plants that they resemble the tallest varieties of the same species. Environmentally limited rosette plants such as spinacea oleraceae, and cabbage generally do't flower in the rossete form. Just before flowering, these plants will undergo extensive internode elongation, a phenomenon known as bolting.

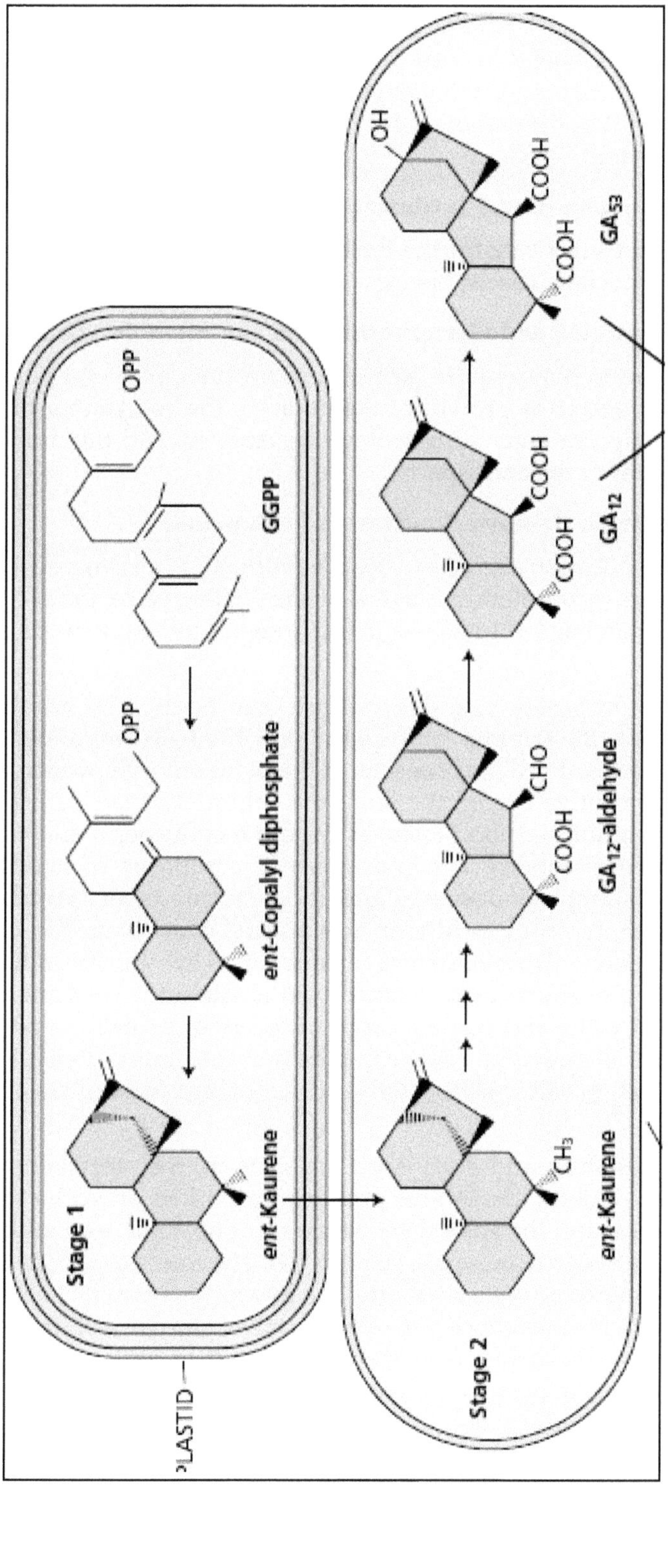

Figure 6.7: Gibberellin Biosynthesis Pathways.

Gibberellins Promote Fruit Set

Applications of gibberellins can cause *fruit set* (the initiation of fruit growth following pollination) and growth of some fruits, in cases where auxin may have no effect. For example, stimulation of fruit set by gibberellin has been observed in apple (*Malus sylvestris*).

Gibberellins Influence Floral Initiation and Sex Determination

Gibberellin can substitute for the longday or cold requirement for flowering in many plants, especially rosette species.

Gibberellins Promote Seed Germination and breaking dormancy

Seed germination may require gibberellins for one of several possible steps: the activation of vegetative growth of the embryo, the weakening of a growth-constraining endosperm layer surrounding the embryo, and the mobilization of stored food reserves of the endosperm.

Gibberelline Induces de-novo Synthesis of α-Amylase

Gibberellin application also stimulates the production of numerous hydrolases, notably ?α-amylase, by the aleurone layers of germinating cereal grains. This aspect of gibberellins action has led to its use in the brewing industry in the production of malt.

Cereal grains (*caryopses;* singular *caryopsis*) can be divided into three parts: the diploid embryo, the triploid endosperm, and the fused testa–pericarp (seed coat–fruit wall). The embryo part consists of the plant embryo proper, along with its specialized absorptive organ, the *scutellum* (plural *scutella*), which functions in absorbing the solubilised food reserves from the endosperm and transmitting them to the growing embryo. The endosperm is composed of two tissues: the centrally located starchy endosperm and the aleurone layer (Figure 5.8). The starchy endosperm, typically nonliving at maturity, consists of thin-walled cells filled with starch grains. The aleurone layer surrounds the starchy endosperm and is cytologically and biochemically distinct from it. Aleurone cells are enclosed in thick primary cell walls and contain large numbers of protein-storing vacuoles called *protein bodies,* enclosed by a single membrane. The protein bodies also contain phytin, a mixed cation salt (mainly Mg^{2+} and K^{+}) of *myo*-inositol hexaphosphoric acid (phytic acid).

During germination and early seedling growth, the stored food reserves of the endosperm–chiefly starch and protein–are broken down by a variety of hydrolytic enzymes, and the solubilized sugars, amino acids, and other products are transported to the growing embryo. The two enzymes responsible for starch degradation are β- and α-amylase. α-Amylase hydrolyzes starch chains internally to produce oligosaccharides consisting of α-1,4-linked glucose residues. α-Amylase degrades these oligosaccharides from the ends to produce maltose, a disaccharide. Maltase then converts maltose to glucose. α-Amylase is secreted into the starchy endosperm of cereal seeds by both the scutellum and the aleurone layer. The sole function of the aleurone layer of the seeds of graminaceous monocots (*e.g.*, barley, wheat, rice, rye, and oats) appears to be the synthesis and release of hydrolytic enzymes. After completing this function, aleurone cells undergo programmed cell death.

The significance of the gibberellin effect became clear when it was shown that the embryo synthesizes and releases gibberellins (chiefly GA1) into the endosperm during germination. Thus the cereal embryo efficiently regulates the mobilization of its own food reserves through the secretion of gibberellins, which stimulate the digestive function of the aleurone layer.

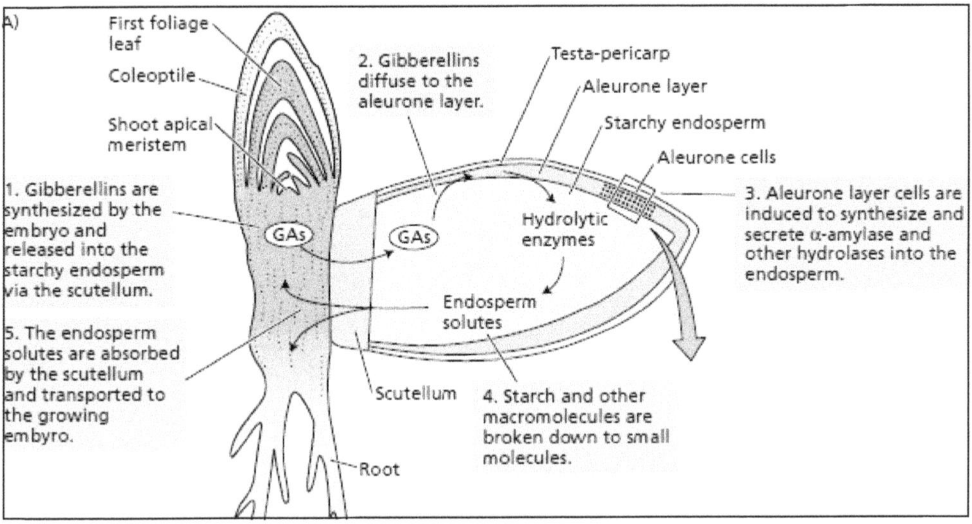

Figure 6.8: Action of Gibberlin in Plants.

Gibberellins have Commercial Applications

The major uses of gibberellins (GA3, unless noted otherwise), applied as a spray or dip, are to manage fruit crops, to malt barley, and to increase sugar yield in sugarcane. In some crops a reduction in height is desirable, and this can be accomplished by the use of gibberellin synthesis inhibitors.

Gibberellins Stimulate Cell Elongation and Cell Division

The effect of gibberellins applied to intact dwarf plants. increases both cell elongation and cell division,as evidenced by increases in cell length and cell number in response to applications of gibberellins.

Gibberellins Enhance Cell Wall Extensibility without Acidification

6.2.4. Gibberellin Signal Transduction

See Figure 6.9.

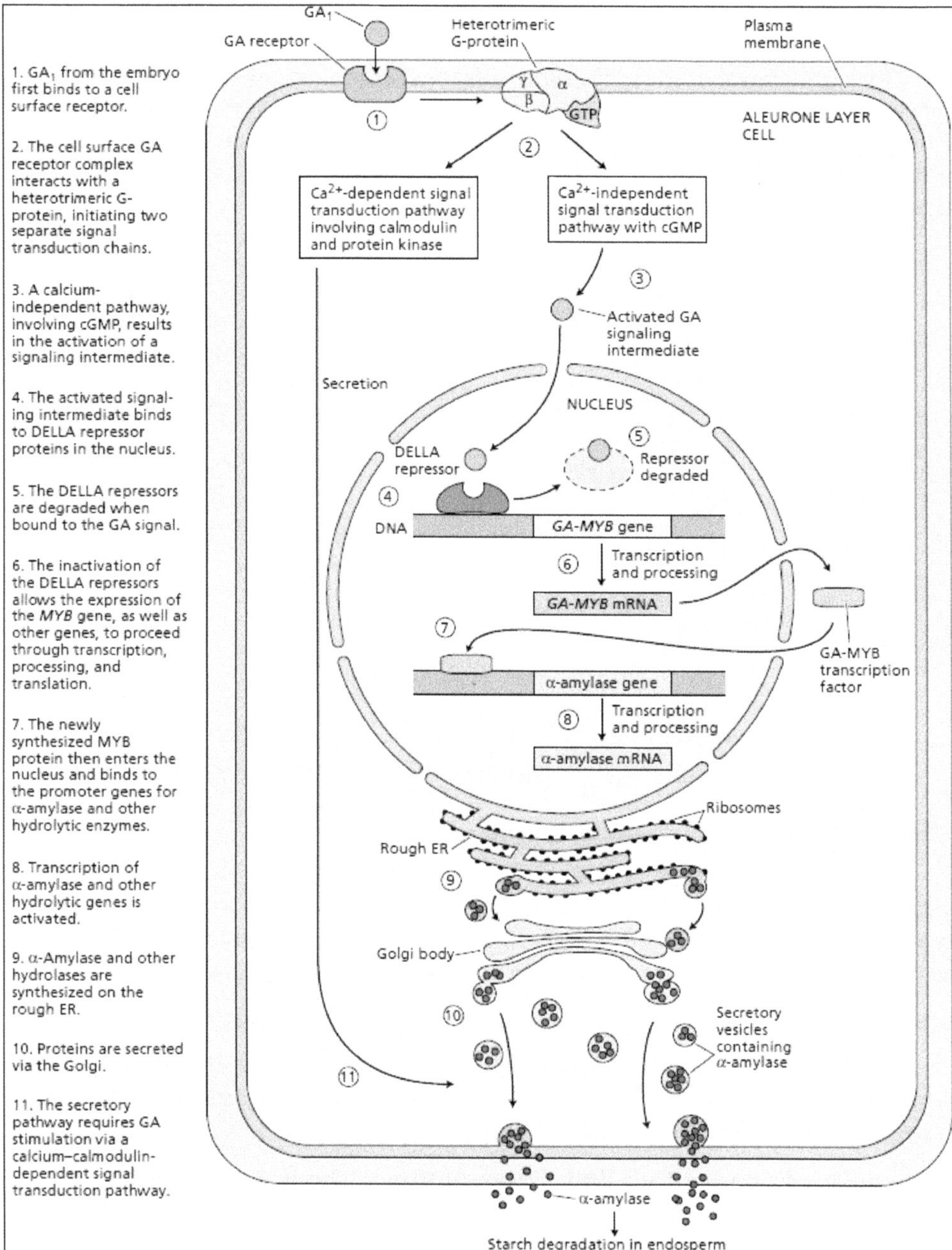

Figure 6.9: Gibberllin Signal Transduction (Courtesy of Taiz and Zieger).

6.3. Cytokinins

A small molecule was identified from the autoclaved DNA and named kinetin. It was shown to be an adenine (or aminopurine) derivative, 6-furfuryl aminopurine (Miller *et al.*, 1955).

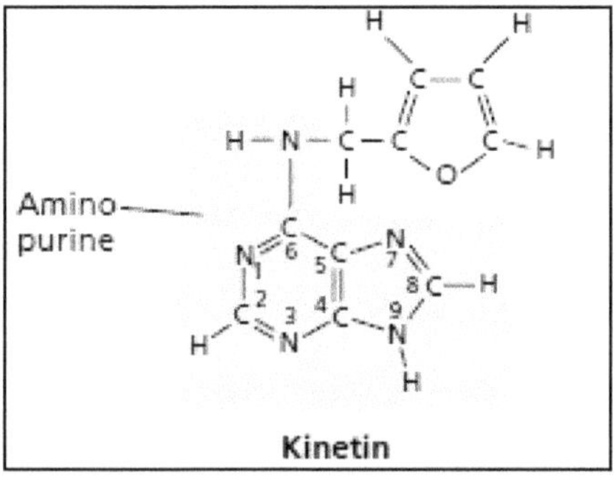

Figure 6.10: Chemical Structure of Cytokinin.

In the presence of an auxin, kinetin would stimulate tobacco pith parenchyma tissue to proliferate in culture. No kinetin-induced cell division occurs without auxin in the culture medium. Kinetin is not a naturally occurring plant growth regulator. It is a by-product of the heat-induced degradation of DNA, in which the deoxyribose sugar of adenosine is converted to a furfuryl ring and shifted from the 9 position to the 6 position on the adenine ring.

Zeatin is the most Abundant Natural Cytokinin

Several years after the discovery of kinetin, extracts of the immature endosperm of corn (*Zea mays*) were found to contain a substance that has the same biological effect as kinetin. This substance stimulates mature plant cells to divide when added to a culture medium along with an auxin. Letham (1973) isolated the molecule responsible for this activity and identified it as *trans*-6-(4-hydroxy-3- methylbut-2-enylamino) purine, which he called zeatin:

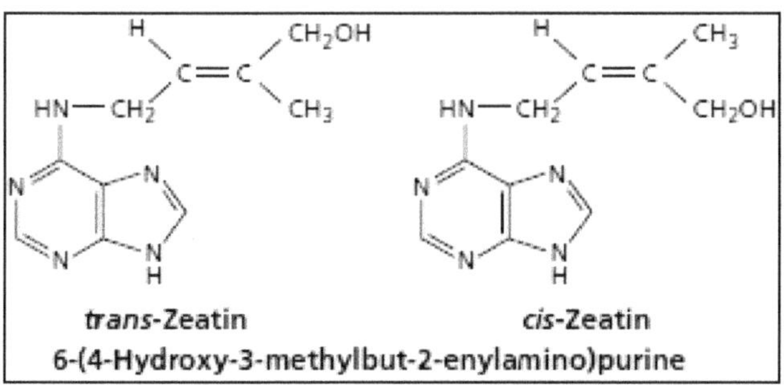

Figure 6.11: Natural Cytokinin.

The molecular structure of zeatin is similar to that of kinetin. Both molecules are adenine or aminopurine derivatives. Although they have different side chains, in both cases the side chain is attached to the 6 nitrogen of the amino purine. Because the side chain of zeatin has a double bond, it can exist in either the *cis* or the *trans* configuration. Free zeatin is mainly trans while zeatin found in t-RNA is in the cis form.

6.3.1. Biosynthesis

A major site of cytokinin biosynthesis in higher plants is the root.High cytokinin levels in roots, especially mititically active root tip and in xylem sap of roots. Immature seed and developing fruits contain high level of cytokinins. Cytokinin

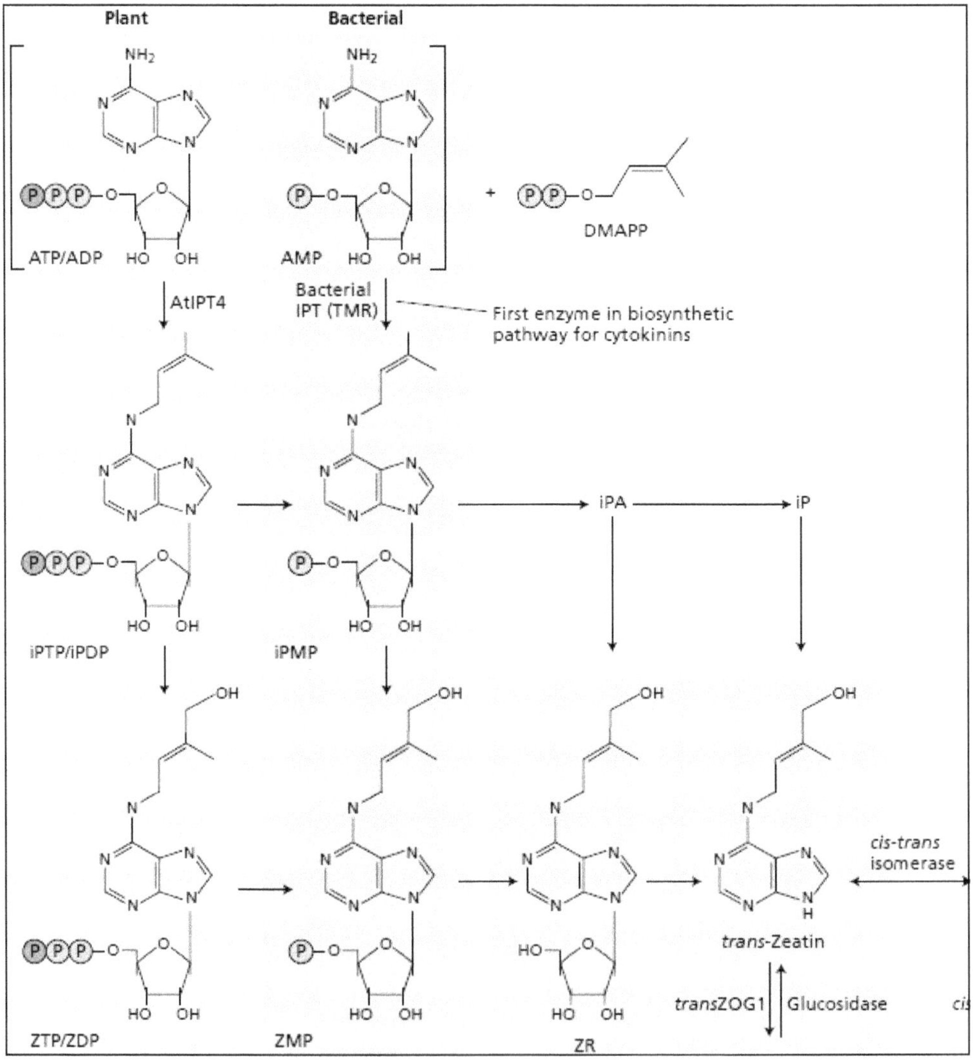

Figure 6.12: Biosynthetic Pathway for Cytokinin Biosynthesis.

are commonly found in the cell as modified bases in transfer ribonucleic acid (t-RNA) molecules.

The first step in cytokinin biosynthesis is the addition of the isopentenyl side chain from DMAPP to an adenosine moiety. The plant and bacterial IPT enzymes differ in the adenosine substrate used; the plant enzyme appears to utilize both ADP and ATP, and the bacterial enzyme utilizes AMP. The products of these reactions (iPMP, iPDP, or iPTP) are converted to zeatin by an unidentified hydroxylase. The various phosphorylated forms can be interconverted and free trans-Zeatin can be formed from the riboside by enzymes of general purine metabolism.

6.3.2. Physiological Effects of Cytokinins

1. Cytokinins Regulate Cell Division in Shoots and Roots

2. Cytokinins Regulate Specific Components of the Cell Cycle

Cytokinins regulate cell division by affecting the controls that govern the passage of the cell through the cell division cycle.

3. The Auxin: Cytokinin Ratio Regulates Morphogenesis in Cultured Tissues

The differentiation of cultured callus tissue derived from tobacco pith segments into either roots or shoots depends on the ratio of auxin to cytokinin in the culture medium. Whereas high auxin:cytokinin ratios stimulated the formation of roots, low auxin:cytokinin ratios led to the formation of shoots. At intermediate levels the tissue grew as an undifferentiated callus. The effect of auxin: cytokinin ratios on morphogenesis can also be seen in crown gall tumors by mutation of the T-DNA of the *Agrobacterium* Ti plasmid.

4. Cytokinins Modify Apical Dominance and Promote Lateral Bud Growth

One of the primary determinants of plant form is the degree of apical dominance. Although apical dominance may be determined primarily by auxin, physiological studies indicate that cytokinins play a role in initiating the growth of lateral buds. For example, direct applications of cytokinins to the axillary buds of many species stimulate cell division activity and growth of the buds. cytokinin-overproducing plants tend to be bushy.

5. Cytokinins Induce Bud Formation in a Moss

The moss *Funaria hygrometrica* is a well-studied example. The germination of moss spores gives rise to a filament of cells called a *protonema* (plural *protonemata*). The protonema elongates and undergoes cell divisions at the tip, and it forms branches some distance back from the tip. Light, especially red light, is required for bud formation in *Funaria*. In the dark, buds fail to develop, but cytokinin added to the medium can substitute for the light requirement. Cytokinin not only stimulates normal bud development; it also increases the total number of buds.

6. Cytokinins Delay Leaf Senescence

Leaves detached from the plant slowly lose chlorophyll, RNA, lipids, and protein, even if they are kept moist and provided with minerals. This programmed aging process leading to death is termed senescence. Leaf senescence is more rapid in the dark than in the light. Treating isolated leaves of many species with cytokinins will delay their senescence. Although applied cytokinins do not prevent senescence completely, their effects can be dramatic, particularly when the cytokinin is sprayed directly on the intact plant. If only one leaf is treated, it remains green after other leaves of similar developmental age have yellowed and dropped off the plant. Even a small spot on a leaf will remain green if treated with a cytokinin, after the surrounding tissues on the same leaf begin to senesce. The resulting elevated cytokinin levels not only blocked senescence, but also limited further expression of the *ipt* gene(senescence promoter gene), preventing cytokinin overproduction. This result suggests that cytokinins are a natural regulator of leaf senescence.

7. Cytokinins Promote Movement of Nutrients

Cytokinins influence the movement of nutrients into leaves from other parts of the plant, a phenomenon known as *cytokinin-induced nutrient mobilization.*

8. Cytokinins Promote Chloroplast Development

Dark-grown seedlings are said to be etiolated. If the etiolated leaves are treated with cytokinin before being illuminated, they form chloroplasts with more extensive grana, and chlorophyll and photosynthetic enzymes are synthesized at a greater rate upon illumination.

9 .Cytokinins Promote Cell Expansion in Leaves and Cotyledons

The promotion of cell enlargement by cytokinins is in the cotyledons of dicots with leafy cotyledons, such as mustard, cucumber, and sunflower. Cytokinin treatment promotes additional cell expansion, with no increase in the dry weight of the treated cotyledons.

Cytokinin-Regulated Processes are Revealed in Plants that Overproduce Cytokinin

The *ipt* gene from the *Agrobacterium* Ti plasmid has been introduced into many species of plants, resulting in cytokinin overproduction. These cytokinin-overproducing plants exhibit several characteristics that point to roles played by cytokinin in plant physiology and development:

- ☆ The shoot apical meristems of cytokinin-overproducing plants produce more leaves.
- ☆ The leaves have higher chlorophyll levels and are much greener.
- ☆ Adventitious shoots may form from unwounded leaf veins and petioles.
- ☆ Leaf senescence is retarded.
- ☆ Apical dominance is greatly reduced.
- ☆ The more extreme cytokinin-overproducing plants are stunted, with greatly shortened internodes.
- ☆ Rooting of stem cuttings is reduced, as is the root growth rate.

6.4. Ethylene

6.4.1. Structure and Biosynthesis of Ethylene

Ethylene can be produced by almost all parts of higher plants.In general, meristematic regions and nodal regions are the most active in ethylene biosynthesis. However, ethylene production also increases during leaf abscission and flower senescence, as well as during fruit ripening. Any type of wounding can induce ethylene biosynthesis, as can physiological stresses such as flooding, chilling, disease, and temperature or drought stress.

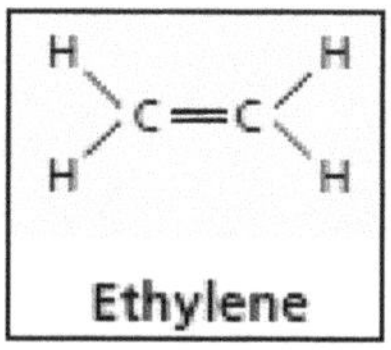

Figure 6.13: Chemical Structure of Ethylene.

Ethylene is the simplest known olefin (its molecular weight is 28), and it is lighter than air under physiological conditions. It is flammable and readily undergoes oxidation. Ethylene can be oxidized to ethylene oxid. Ethylene is released easily from the tissue and diffuses in the gas phase through the intercellular spaces and outside the tissue.

In most plant tissues, ethylene can be completely oxidized to CO_2, in the following reaction:

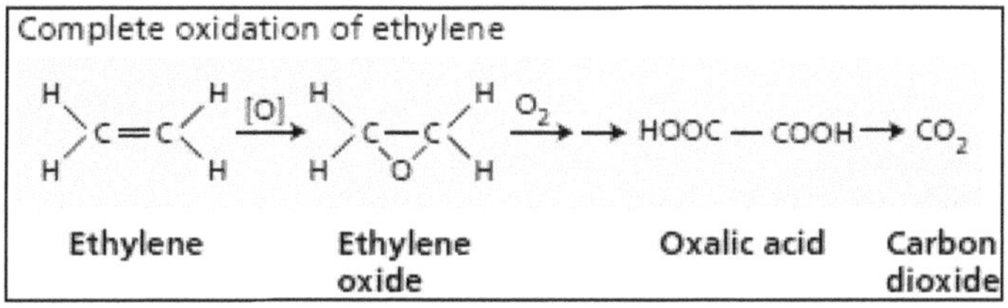

All organs of higher plants can synthesize ethylene. Ethylene is biologically active at very low concentrations–less than 1 part per million.

S-adenosylmethionine (AdoMet), which is synthesized from methionine and ATP, is an intermediate in the ethylene biosynthetic pathway, and the immediate precursor of ethylene is 1-aminocyclopropane-1-carboxylic acid (ACC). The role of ACC became evident in experiments in which plants were treated with [14C] methionine. Under anaerobic conditions, ethylene was not produced from the [14C] methionine, and labeled ACC accumulated in the tissue. On exposure to oxygen, however, ethylene production surged. The labeled ACC was rapidly converted to ethylene in the presence of oxygen by various plant tissues, suggesting that ACC is the immediate precursor of ethylene in higher plants and that oxygen is required for the conversion. In general, when ACC is supplied exogenously to plant tissues, ethylene production increases substantially.

ACC synthase is an unstable, cytosolic enzyme. Its level is regulated by environmental and internal factors, such as wounding, drought stress, flooding, and auxin. Because ACC synthase is present in such low amounts in plant tissues (0.0001 per cent of the total protein of ripe tomato).

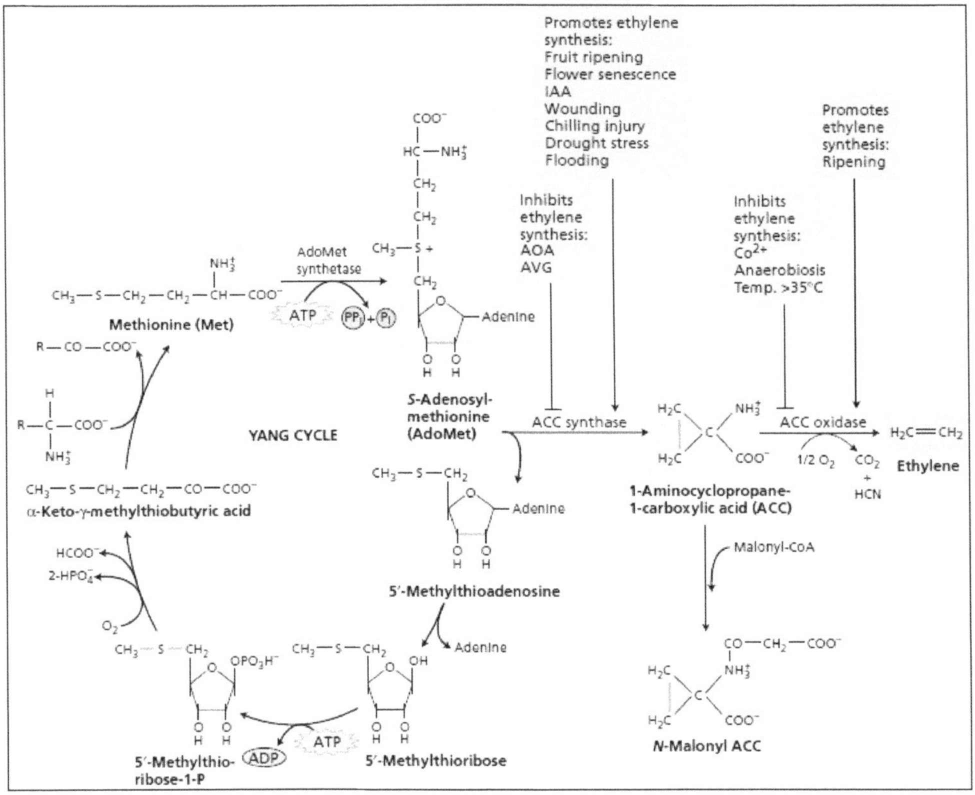

Figure 6.14: Ethylene Biosynthesis Pathway.

ACC oxidase catalyzes the last step in ethylene biosynthesis: the conversion of ACC to ethyleneIn tissues that show high rates of ethylene production, such as ripening fruit, ACC oxidase activity can be the rate-limiting step in ethylene biosynthesis.

6.4.2. Physiological Effects of Ethylene

Ethylene Promotes the Ripening of Some Fruits

Ethylene has long been recognized as the hormone that accelerates the ripening of edible fruits. Exposure of such fruits to ethylene hastens the processes associated with ripening, and a dramatic increase in ethylene production accompanies the initiation of ripening.

Leaf Epinasty Results when ACC from the Root is Transported to the Shoot

The downward curvature of leaves that occurs when the upper (adaxial) side of the petiole grows faster than the lower (abaxial) side is termed epinasty. Ethylene and high concentrations of auxin induce epinasty, and it has now been established that auxin acts indirectly by inducing ethylene production.

Ethylene Induces Lateral Cell Expansion

At concentrations above 0.1 L L–1, ethylene changes the growth pattern of seedlings by reducing the rate of elongation and increasing lateral expansion, leading to swelling of the region below the hook. These effects of ethylene are common to growing shoots of most dicots, forming part of the triple response. The triple response consists of inhibition and swelling of the hypocotyl, inhibition of root elongation, and exaggeration of the apical hook.

The Hooks of Dark-Grown Seedlings are Maintained by Ethylene Production

Etiolated dicot seedlings are usually characterized by a pronounced hook located just behind the shoot apex.

Ethylene Breaks Seed and Bud Dormancy in Some Species

Ethylene has the ability to break dormancy and initiate germination in certain seeds, such as cereals. In addition to its effect on dormancy, ethylene increases the rate of seed germination of several species.

Ethylene Induces the Formation of Roots and Root Hairs

Ethylene is capable of inducing adventitious root formation in leaves, stems, flower stems, and even other roots.

Ethylene has also been shown to act as a positive regulator of root hair formation in several species.

Ethylene Induces Flowering in the Pineapple Family

Although ethylene inhibits flowering in many species, it induces flowering in pineapple and its relatives.

Ethylene Enhances the Rate of Leaf Senescence

- ☆ Exogenous applications of ethylene or ACC (the precursor of ethylene) accelerate leaf senescence, and treatment with exogenous cytokinins delays leaf senescence.
- ☆ Enhanced ethylene production is associated with chlorophyll loss and color fading, which are characteristic features of leaf and flower senescence,an inverse correlation has been found between cytokinin levels in leaves and the onset of senescence.
- ☆ Inhibitors of ethylene synthesis (*e.g.*, AVG or CO_2^+) and action (*e.g.*, Ag^+ or CO_2) retard leaf senescence.

Senescence is regulated by the balance of ethylene and cytokinin. In addition, abscisic acid (ABA) has been implicated in the control of leaf senescence.

6.4.3. Ethylene Biosynthesis in the Abscission Zone is Regulated by Auxin

The shedding of leaves, fruits, flowers, and other plant organs is termed abscission. Abscission takes place in specific layers of cells, called abscission

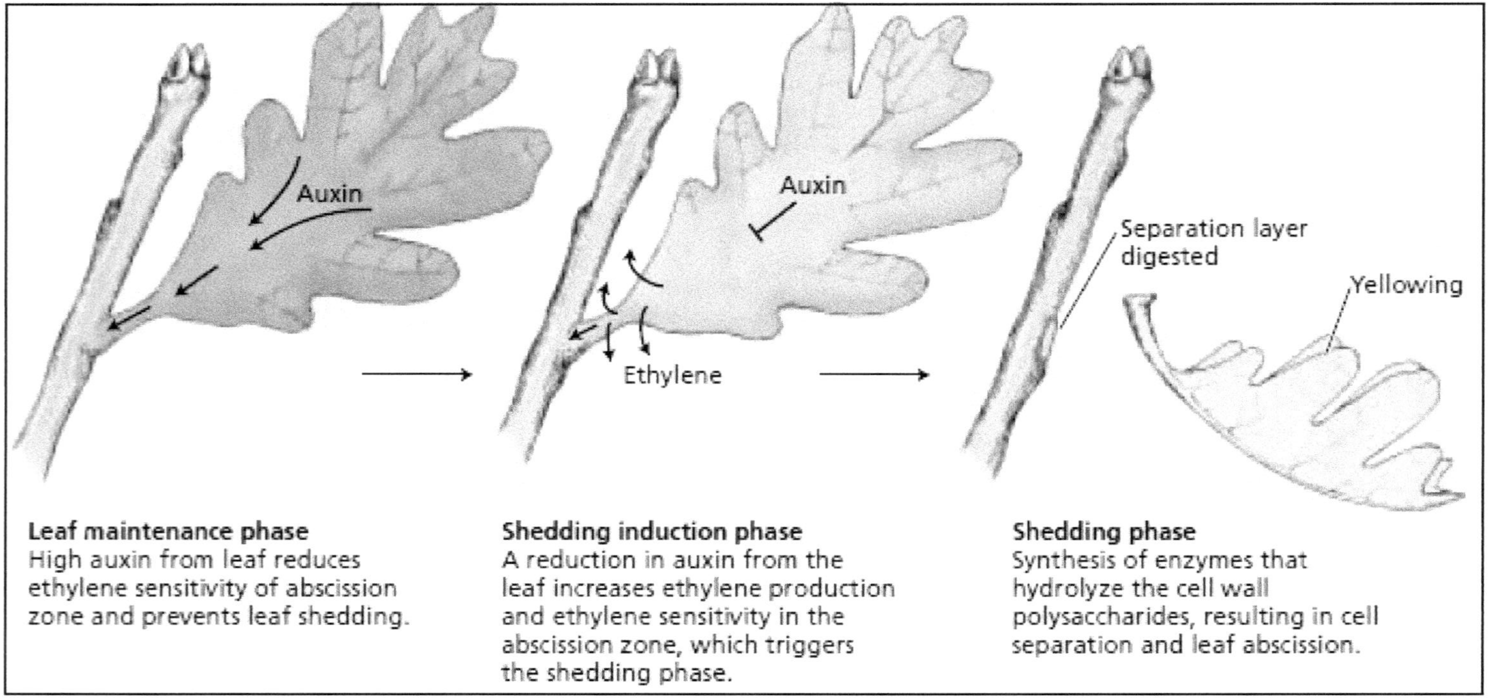

Figure 6.15: The Roles of Auxin and Ethylene during Leaf Abscission.

layers, which become morphologically and biochemically differentiated during organ development. Weakening of the cell walls at the abscission layer depends on cell wall–degrading enzymes such as cellulase and polygalacturonase. Ethylene appears to be the primary regulator of the abscission process, with auxin acting as a suppressor of the ethylene effect. However,supraoptimal auxin concentrations stimulate ethylene production, which has led to the use of auxin analogs as defoliants. For example, 2,4,5-T, the active ingredient in Agent Orange, was widely used as a defoliant during the Vietnam War. Its action is based on its ability to increase ethylene biosynthesis, thereby stimulating leaf abscission.

6.4.4. Ethylene has Important Commercial Uses

It is one of the most widely used plant hormones in agriculture. Auxins and ACC can trigger the natural biosynthesis of ethylene and in several cases are used in agricultural practice. Because of its high diffusion rate, ethylene is very difficult to apply in the field as a gas, but this limitation can be overcome if an ethylene-releasing compound is used. The most widely used such compound is ethephon, or 2- chloroethylphosphonic acid.

6.5. Abscisic Acid (ABA)

Abscisic acid has been found to be a ubiquitous plant hormone in vascular plants. It has been detected in mosses but appears to be absent in liverworts. Several genera of fungi make ABA as a secondary metabolite. Within the plant, ABA has been detected in every major organ or living tissue from the root cap to the apical bud. ABA is synthesized in almost all cells that contain chloroplasts or amyloplasts.

6.5.1. Biosynthesis and Metabolism

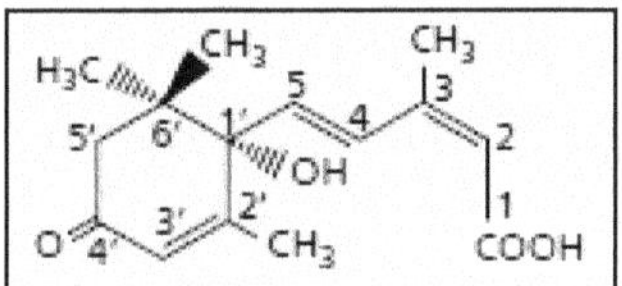

(S)-*cis*-ABA
(naturally occurring active form)

ABA biosynthesis takes place in chloroplasts and other plastids. The pathway begins with isopentenyl diphosphate (IPP), the biological isoprene unit, and leads to the synthesis of the C40 xanthophyll (*i.e.*, oxygenated carotenoid) violaxanthin (Figure 6.16). Synthesis of violaxanthin is catalyzed by zeaxanthin epoxidase (ZEP). Violaxanthin is converted to the C40 compound 9-*cis*-neoxanthin, which is then cleaved to form the C15 compound-xanthoxal, previously called *xanthoxin*, a neutral growth inhibitor that has physiological properties similar to those of ABA. The cleavage is catalyzed by 9-*cis*-epoxycarotenoid dioxygenase (NCED).

Finally, xanthoxal is converted to ABA via oxidative steps involving the intermediate(s) ABA-aldehyde and/or possibly xanthoxic acid. This final step is catalyzed by a family of aldehyde oxidases that all require a molybdenum cofactor.

ABA levels can increase 100-fold within a few days and then decline to vanishingly low levels as maturation proceeds. Under conditions of water stress, ABA in the leaves can increase 50-fold within 4 to 8 hours. Upon rewatering, the ABAlevel declines to normal in the same amount of time. ABAis transported by both the xylem and the phloem, but it is normally much more abundant in the phloem

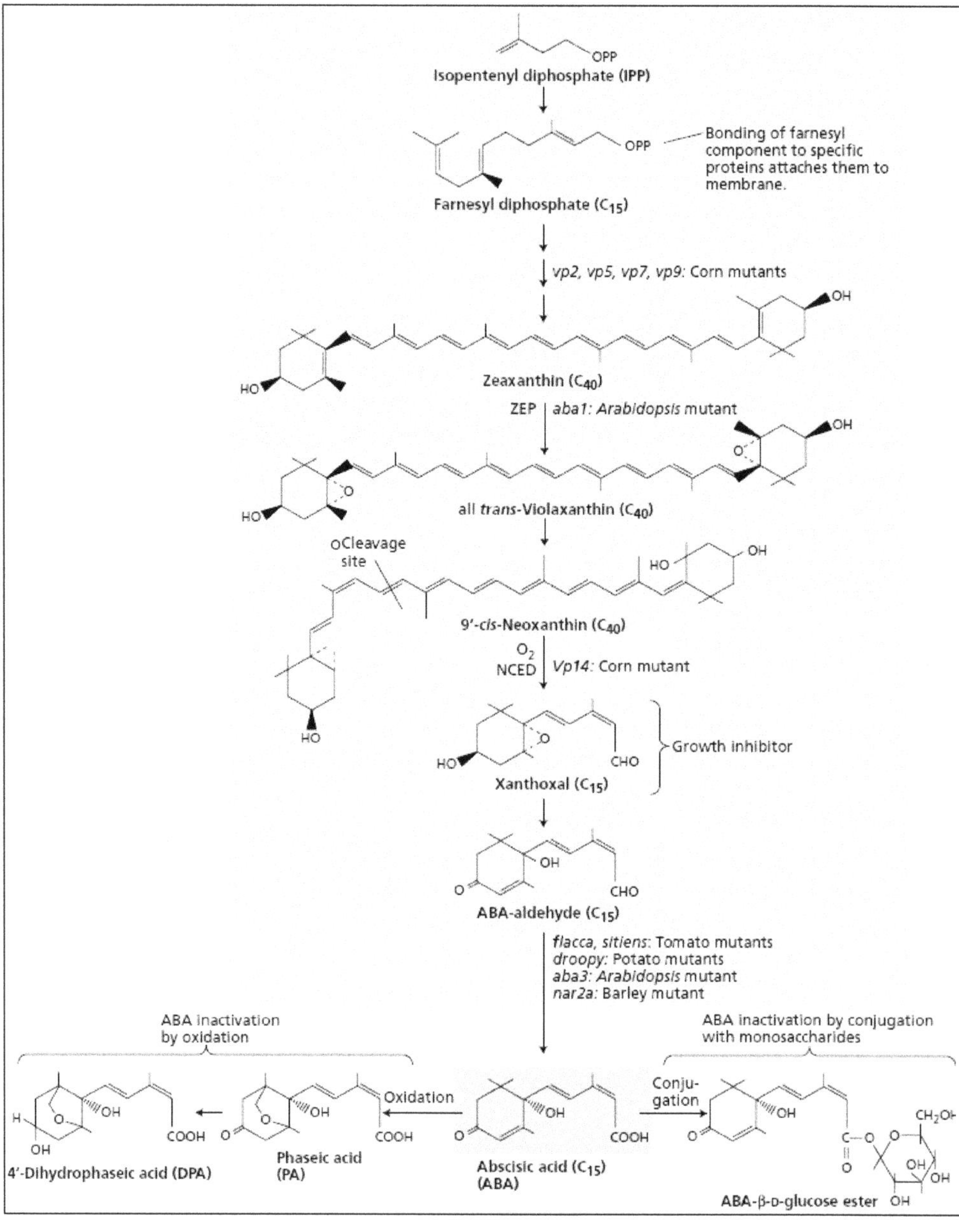

Figure 6.16: Biosynthesis of ABA (Courtesy of Taiz and Zieger).

sap. The inactivation of free ABA is oxidation, yielding the unstable intermediate 6-hydroxymethyl ABA,which is rapidly converted to phaseic acid (PA) and dihydrophaseic acid (DPA). Free ABA is also inactivated by covalent conjugation to another molecule, such as a monosaccharide. A common example of an ABA conjugate is ABA-b-D-glucosyl ester (ABA-GE).

6.5.2. Physiological Effects of ABA

Abscisic acid plays primary regulatory roles in the initiation and maintenance of seed and bud dormancy and in the plant's response to stress, particularly water stress. The diverse physiological effects of ABA, beginning with its role in seed development.

ABA Promotes Desiccation Tolerance in the Embryo

An important function of ABA in the developing seed is to promote the acquisition of desiccation tolerance. During the mid- to late stages of seed development, specific mRNAs accumulate in embryos at the time of high levels of endogenous ABA. These mRNAs encode so-called late-embryogenesis-abundant (LEA) proteins thought to be involved in desiccation tolerance. Synthesis of many LEA proteins, or related family members, can be induced by ABA treatment of either young embryos or vegetative tissues. Thus the synthesis of most LEA proteins is under ABA control. Storage compounds accumulate during mid- to late embryogenesis. Because ABA levels are still high.

ABA Inhibits Precocious Germination and Vivipary

When immature embryos are removed from their seeds and placed in culture midway through development before the onset of dormancy, they germinate precociously–that is, without passing through the normal quiescent and/or dormant stage of development. ABA added to the culture medium inhibits precocious germination.

ABA Accumulates in Dormant Buds

ABA is known as the dormancy-inducing hormone because it accumulates in dormant buds and decreases after the tissue is exposed to low temperatures.

ABA Inhibits GA-Induced Enzyme Production

ABA inhibits the synthesis of hydrolytic enzymes that are essential for the breakdown of storage reserves in seeds.

ABA Closes Stomata in Response to Water Stress

Biosynthesis of ABA is very effective in causing stomatal closure, and its accumulation in stressed leaves plays an important role in the reduction of water loss by transpiration under water stress conditions

ABA Promotes Leaf Senescence Independently of Ethylene

Abscisic acid was originally isolated as an abscission-causing factor. However, it has since become evident that ABA stimulates abscission of organs in only a few species and that the primary hormone causing abscission is ethylene. On the other hand, ABA is clearly involved in leaf senescence, and through its promotion of senescence it might indirectly increase ethylene formation and stimulate abscission.

References

Badescu, G.O. and Napier, R.M. (2006). Receptors for auxin: Will it end in TIRs? *Trends in Plant Sciences* 11: 217-223.

Blakeslee, J.J., Peer, W.A. and Murphy, A.S. (2005). Auxin transport. *Current Opinion in Plant Biology*, 8: 494-500.

Buchanan, B.B., Gruissm, W. and Jones, R.L. (2000). *Biochemistry and Molecular Biology of Plants*. Rockville, MD: American Society of Plant Physiologosts.

Eckardt, N.A. (2007). GA perception and signal transduction: Molecular inte raction of the GA receptor GIDI with GA and the DELLA protein SLR1 in Rice. *The Plant Cell*, 19: 2095-2097.

Chapter 7

Secondary Metabolites

Plants produce a large, diverse array of organic compounds that appear to have no direct function in growth and development. These substances are known as secondary metabolites, *secondary products*, or *natural products*. Secondary metabolites have no generally recognized, direct roles in the processes of photosynthesis, respiration, solute transport, translocation, protein synthesis, nutrient assimilation, differentiation, or the formation of carbohydrates, proteins, and lipids. Secondary metabolites also differ from primary metabolites (amino acids, nucleotides, sugars, acyl lipids) in having a restricted distribution in the plant kingdom. That is, particular secondary metabolites are often found in only one plant species or related group of species, whereas primary metabolites are found throughout the plant kingdom.

Plant secondary metabolites can be divided into three chemically distinct groups:

1. Terpenes,
2. Phenolics, and
3. Nitrogen-containing compounds

7.1. TERPENES

The terpenes, or *terpenoids*, constitute the largest class of secondary products. They are generally insoluble in water. They are biosynthesized from acetyl-CoA or glycolytic intermediates.

All terpenes are derived from the union of five-carbon elements that have the branched carbon skeleton of isopentane:

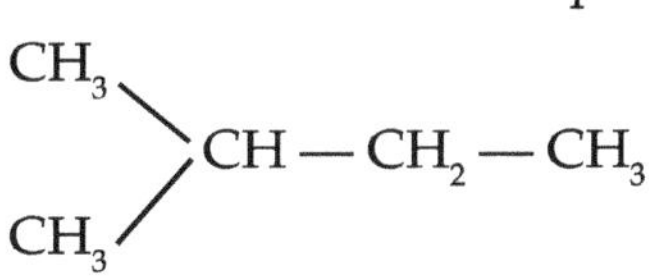

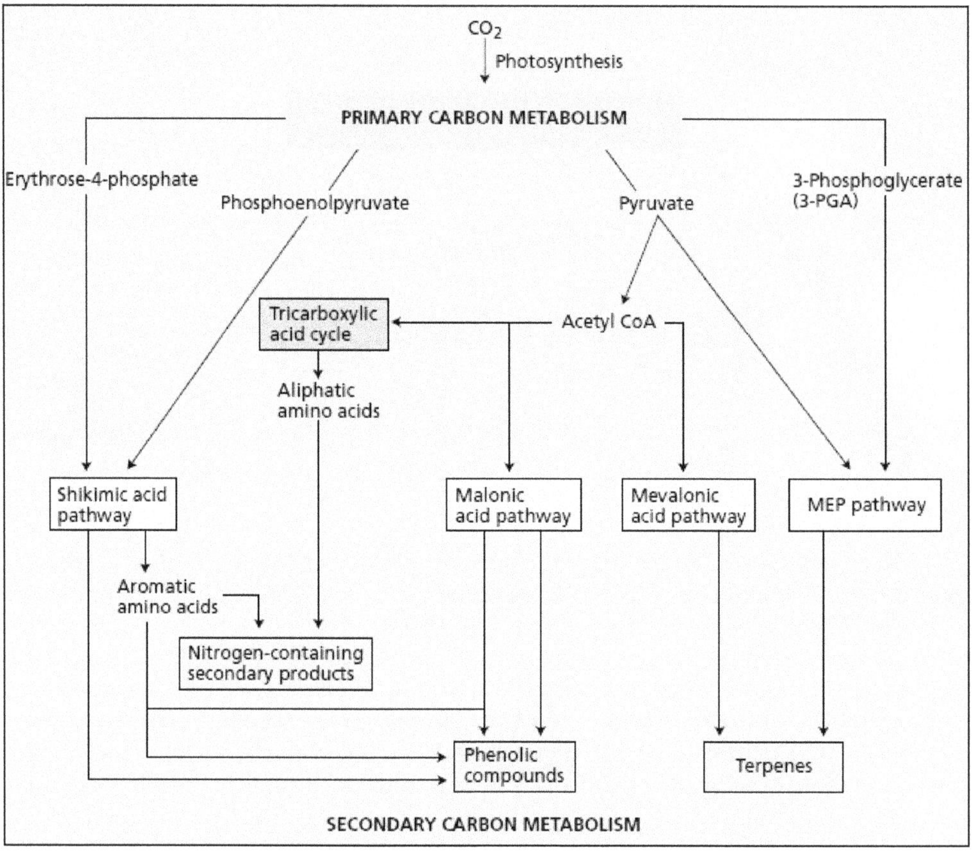

Figure 7.1: A Simplified View of the Major Pathways of Secondary-metabolite Biosynthesis.

The basic structural elements of terpenes are sometimes called isoprene units because terpenes can decompose at high temperatures to give isoprene:

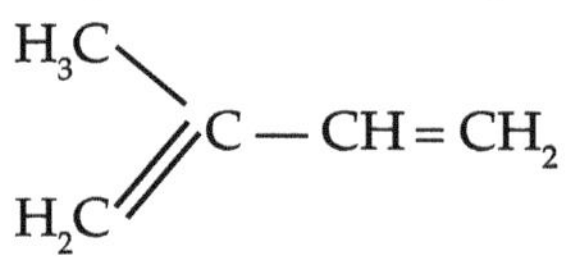

Thus all terpenes are occasionally referred to as *isoprenoids*. Terpenes are classified by the number of five-carbon units they contain. Ten-carbon terpenes, which contain two C5 units, are called *monoterpenes*; 15-carbon terpenes (three C5 units) are *sesquiterpenes*; and 20-carbon terpenes (four C5 units) are *diterpenes*. Larger terpenes include *triterpenes* (30 carbons), *tetraterpenes* (40 carbons), and *polyterpenoids* ([C5]n carbons, where $n > 8$).

Terpene Biosynthesis

Terpenes are biosynthesized from primary metabolites by mevalonic acid pathway, three molecules of acetyl-CoA are joined together stepwise to form

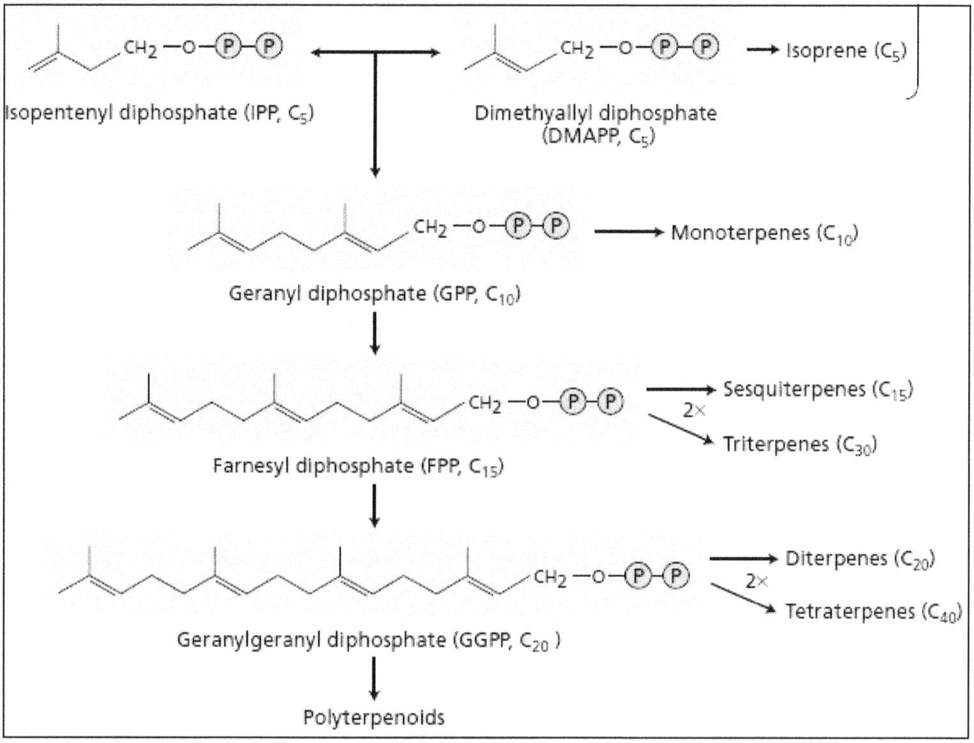

Figure 7.2: Biosynthesis of Terpenes by Mevalonic Acid Pathway.

mevalonic acid. This key six-carbon intermediate is then pyrophosphorylated, decarboxylated, and dehydrated to yield isopentenyl diphosphate (IPP2). IPP is the activated five-carbon building block of terpenes. Recently, it was discovered that IPP also can be formed from intermediates of glycolysis or the photosynthetic carbon reduction cycle.

Isopentenyl diphosphate and its isomer, dimethylallyl diphosphate (DPP), are the activated five-carbon building blocks of terpene biosynthesis that join together to form larger molecules. First IPP and DPP react to give geranyl diphosphate (GPP), the 10-carbon precursor of nearly all the monoterpenes. GPP can then link to another molecule of IPP to give the 15-carbon compound farnesyl diphosphate (FPP), the precursor of nearly all the sesquiterpenes. Addition of yet another molecule of IPP gives the 20-carbon compound geranylgeranyl diphosphate (GGPP), the precursor of the diterpenes. Finally, FPP and GGPP can dimerize to give the triterpenes (C30) and the tetraterpenes (C40), respectively.

Some Terpenes have Roles in Growth and Development

Certain terpenes have a well-characterized function in plant growth or development and so can be considered primary rather than secondary metabolites. For example, the gibberellins, an important group of plant hormones, are diterpenes. Sterols are triterpene derivatives that are essential components of cell membranes, which they stabilize by interacting with phospholipids. The red, orange, and yellow carotenoids are tetraterpenes that function as accessory pigments in photosynthesis and protect photosynthetic tissues from photooxidation. The hormone abscisic acid is a C15 terpene produced by degradation of a carotenoid precursor. Long-chain polyterpene alcohols known as ***dolichols*** function as carriers of sugars in cell wall and glycoprotein synthesis.Terpene-derived side chains, such as the phytol side chain of chlorophyll, help anchor certain molecules in membranes. Thus various terpenes have important primary roles in plants. However, the vast majority of the different terpene structures produced by plants are secondary metabolites that are presumed to be involved in defense.

Terpenes Defend against Herbivores in many Plants

Terpenes are toxins and feeding deterrents to many plantfeeding insects and mammals; thus they appear to play important defensive roles in the plant kingdom (Gershenzon and Croteau, 1992). For example, the monoterpene esters called pyrethroids that occur in the leaves and flowers of *Chrysanthemum* species show very striking insecticidal activity. In conifers such as pine, monoterpenes accumulate in resin ducts found in the needles, twigs, and trunk. These compounds are toxic to numerous insects, including bark beetles, which are serious pests of conifer species throughout the world.

Many plants contain mixtures of volatile monoterpenes and sesquiterpenes, called essential oils. Peppermint, lemon, basil, and sage are examples of plants that contain Essential oils. Essential oil have well-known insect repellent properties. The chief monoterpene constituent of peppermint oil is menthol; that of lemon oil is limonene.

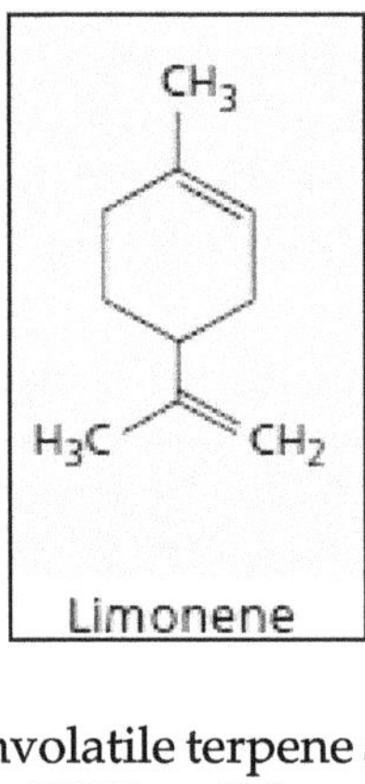

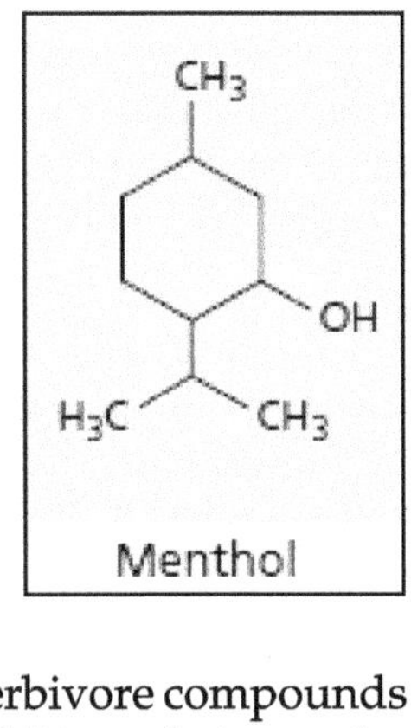

Among the nonvolatile terpene antiherbivore compounds are the limonoids, a group of triterpenes (C30) well known as bitter substances in citrus fruit. Perhaps the most powerful deterrent to insect feeding known is *azadirachtin*, a complex limonoid from the neem tree (*Azadirachta indica*).

The phytoecdysones, first isolated from the common fern, *Polypodium vulgare*, are a group of plant steroids that have the same basic structure as insect molting hormones. Ingestion of phytoecdysones by insects disrupts molting and other developmental processes, often with lethal consequences. Triterpenes that are active against vertebrate herbivores include cardenolides and saponins. Cardenolides are glycosides (compounds containing an attached sugar or sugars) tha t taste bitter and are extremely toxic to higher animals. Cardenolides extracted from species of foxglove (*Digitalis*) are used for the treatment of heart disease.

Saponins are steroid and triterpene glycosides, so named because of their soaplike properties. The presence of both lipid-soluble (the steroid or triterpene) and watersoluble (the sugar) elements in one molecule gives saponins detergent properties, and they form a soapy lather when shaken with water.

7.2. Phenolic Compounds

Plants produce a large variety of secondary products that contain a phenol group–a hydroxyl functional group on an aromatic ring. Many phenolic compounds serve as defense compounds against herbivores and pathogens. Others function in mechanical support, in attracting pollinators and fruit dispersers, in absorbing harmful ultraviolet radiation, or in reducing the growth of nearby competing plants (allelopathy).

Biosynthesis of most Plant Phenolics

The shikimic acid pathway participates in the biosynthesis of most plant phenolics. The shikimic acid pathway converts simple carbohydrate precursors derived from glycolysis and the pentose phosphate pathway to the aromatic amino acids (Figures 7.3 and 7.4).

7.3. Lignin

After cellulose, the most abundant organic substance in plants is lignin, a highly branched polymer of phenylpropanoid groups that plays both primary and secondary roles. Lignin is generally formed from three different phenylpropanoid alcohols: coniferyl, coumaryl, and sinapyl, alcohols which are synthesized from phenylalanine via various cinnamic acid derivatives. Lignin is found in the cell walls of various types of supporting and conducting tissue, notably the tracheids and vessel elements of the xylem. It is deposited chiefly in the thickened secondary wall but can also occur in the primary wall and middle lamella in close contact with the celluloses and hemicelluloses already present. The mechanical rigidity of lignin strengthens stems and vascular tissue, allowing upward growth and permitting water and minerals to be conducted through the xylem under negative pressure without collapse of the tissue and lignin has significant protective functions in plants.

7.4. Flavonoids

The basic carbon skeleton of a flavonoid contains 15 carbons arranged in two aromatic rings connected by a three-carbon bridge:

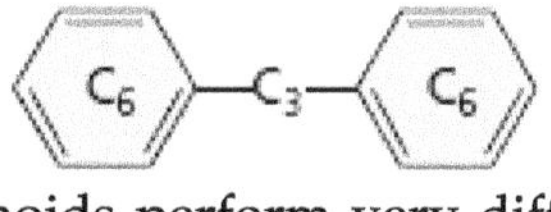

Different types of flavonoids perform very different functions in the plant, including pigmentation and defense.There Are Four Major Groups of Flavonoids - Anthocyanins, flavones, Flavonols, and Isoflavonoids.

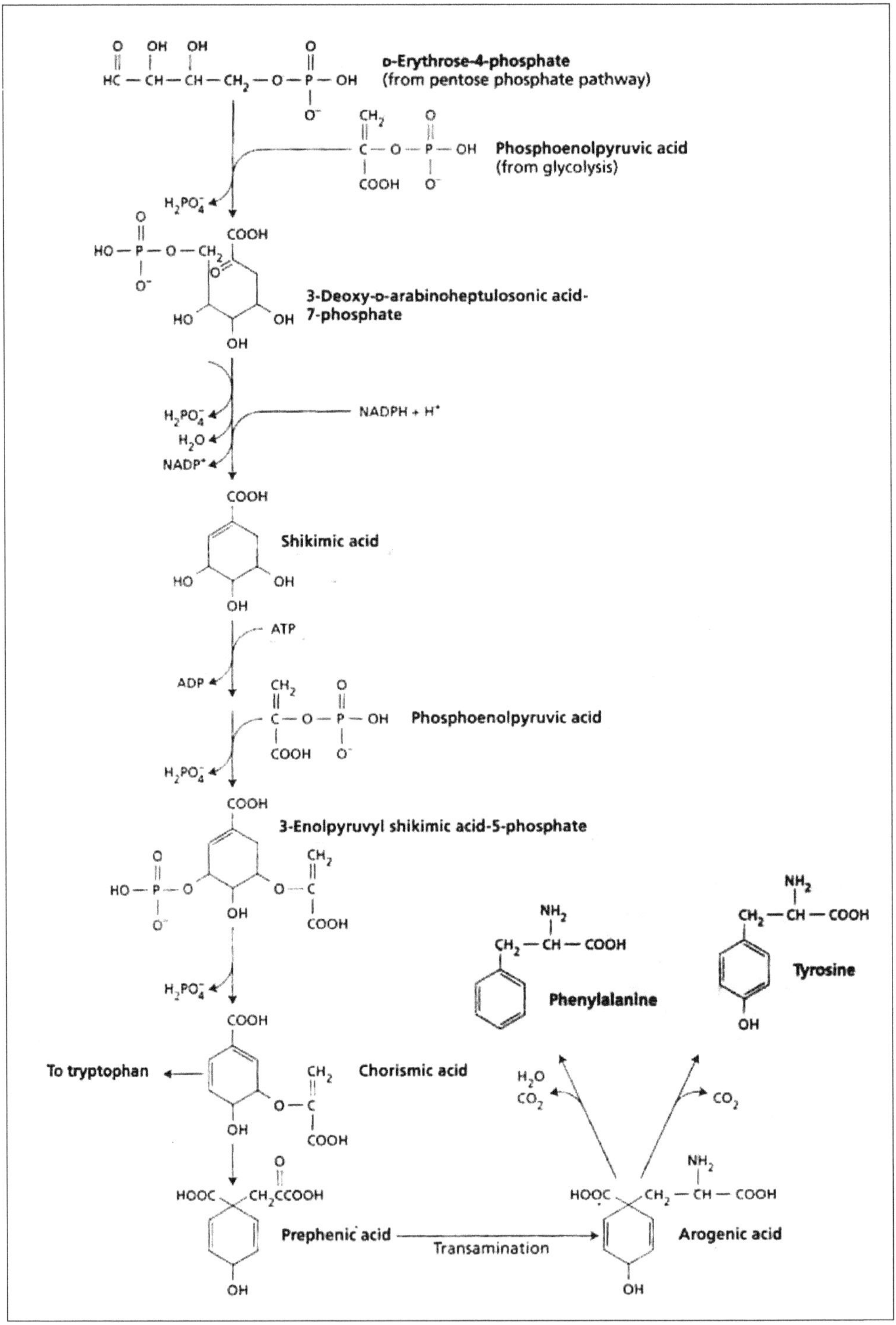

Figure 7.3: Shikimic Acid Pathway.

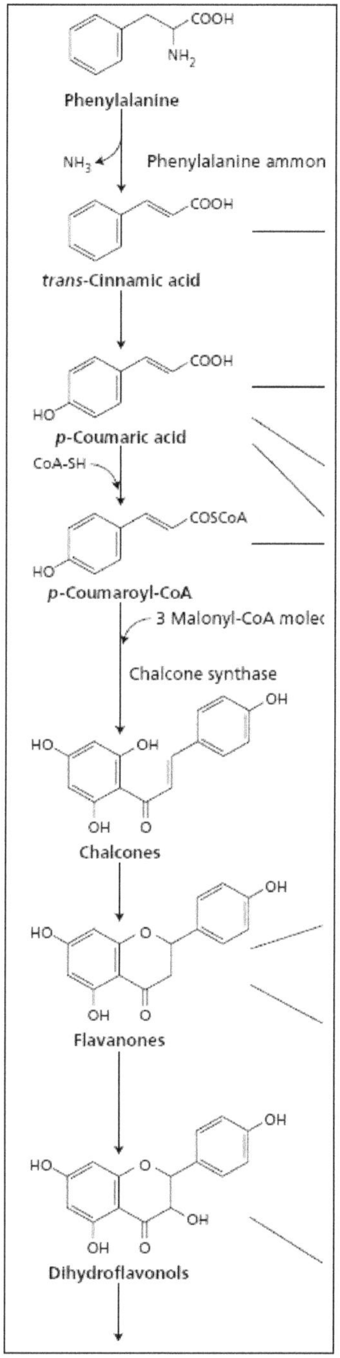

Figure 7.4: Outline of Phenolic Biosynthesis from Phenylalanine.

The formation of many plant phenolics including simple phenylpropanoids, coumarins, benzoic acid derivatives, lignin, anthocyanins, isoflavones, condensed tannins and other flavonoids.

The colored pigments of plants are of two principal types: carotenoids and flavonoids. The most widespread group of pigmented flavonoids is the anthocyanins, which are responsible for most of the red,pink, purple, and blue colors observed in plant parts. By coloring flowers and fruits, the anthocyanins are vitally important in attracting animals for pollination and seed dispersal.

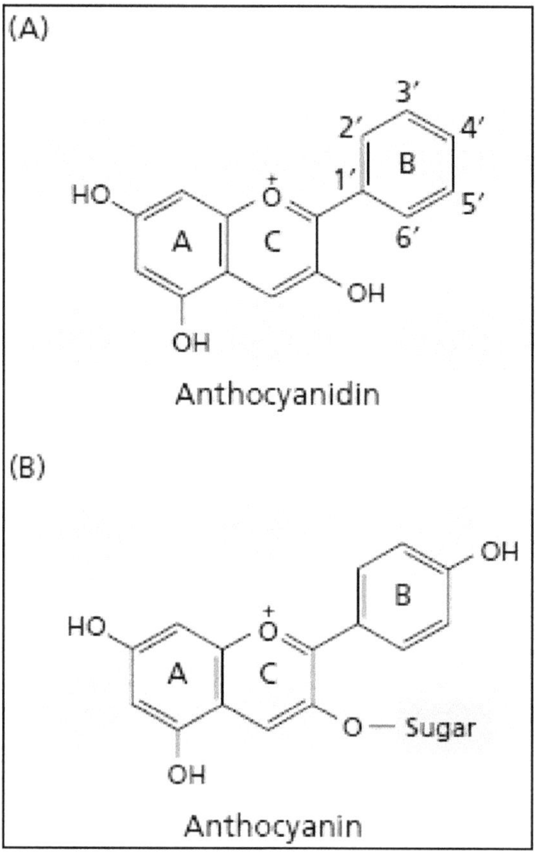

Figure 7.5: Structure of Anthocyanin and Anthocyanidin.

Anthocyanins may also exist in supramolecular complexes along with chelated metal ions and flavone copigments. The blue pigment of dayflower (*Commelina communis*) was found to consist of a large complex of six anthocyanin molecules.

Two other major groups of flavonoids found in flowers are flavones and flavonols. These flavonoids generally absorb light at shorter wavelengths than anthocyanins do, so they are not visible to the human eye. However, insects such as bees, which see farther into the ultraviolet range of the spectrum than humans do, may respond to flavones and flavonols as attractant cues. Flavones and flavonols are also present in the leaves of all green plants. These two classes of flavonoids function to protect cells from excessive UV-B radiation (280–320 nm). flavones and flavonols secreted into the soil by legume roots mediate the interaction of legumes and nitrogen-fixing symbionts. flavonoids also play a regulatory role in plant development as modulators of polar auxin transport.

Isoflavonoids are found mostly in legumes and have several different biological activities. Some, such as the rotenoids, have strong insecticidal actions; others have anti-estrogenic effects. For example, sheep grazing on clover rich in isoflavonoids often suffer from infertility. Isoflavonoids have best known for their role as phytoalexins, antimicrobial compounds synthesized in response to bacterial or fungal infection that help limit the spread of the invading pathogen.

7.5. Tannins

A second category of plant phenolic polymers with defensive properties, besides lignins, is the tannins. Tannins bind the collagen proteins of animal hides, increasing their resistance to heat, water, and microbes. There are two categories of tannins: condensed and hydrolyzable. Condensed tannins are compounds formed by the polymerization of flavonoid units. Interestingly, humans often prefer a certain level of astringency in tannin-containing foods, such as apples, blackberries, tea, and red wine. Plant tannins also serve as defenses against microorganisms. For example, the nonliving heartwood of many trees contains high concentrations of tannins that help prevent fungal and bacterial decay.

7.6. Nitrogen-Containing Compounds

The various nitrogen-containing secondary metabolites, including alkaloids, cyanogenic glycosides, glucosinolates, and nonprotein amino acids.

7.6.1. Alkaloids

The alkaloids are a large family of more than 15,000 nitrogen- containing secondary metabolites found in approximately 20 per cent of the species of vascular plants. Alkaloids are positively charged and are generally water soluble. Alkaloids are usually synthesized from one of a few common amino acids–in particular, lysine, tyrosine, and tryptophan. Most alkaloids are now believed to function as defenses against predators, especially mammals. Nearly all alkaloids are also toxic to humans when taken in sufficient quantity. For example, strychnine, atropine, and coniine (from poison hemlock) are classic alkaloid poisoning agents. At lower doses, however, many are useful pharmacologically. Morphine, codeine, and scopolamine are just a few of the plant alkaloids currently used in medicine. Other alkaloids, including cocaine, nicotine, and caffeine enjoy widespread nonmedical use as stimulants or sedatives. On a cellular level, many alkaloids interfere with components of the nervous system, especially the chemical transmitters; others affect membrane transport, protein synthesis, or miscellaneous enzyme activities. Large numbers of livestock deaths are caused by the ingestion of alkaloid-containing plants. such as lupines (*Lupinus*), larkspur (*Delphinium*), and groundsel (*Senecio*) (Figure 7.6)

Alkaloid class	Structure	Biosynthetic precursor	Examples	Human uses
Pyrrolidine		Ornithine (aspartate)	Nicotine	Stimulant, depressant, tranquilizer
Tropane	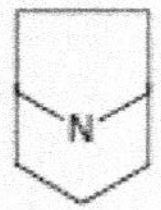	Ornithine	Atropine	Prevention of intestinal spasms, antidote to other poisons, dilation of pupils for examination
			Cocaine	Stimulant of the central nervous system, local anesthetic
Piperidine	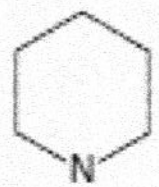	Lysine (or acetate)	Coniine	Poison (paralyzes motor neurons)
Pyrrolizidine		Ornithine	Retrorsine	None
Quinolizidine	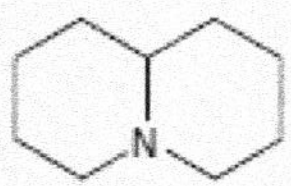	Lysine	Lupinine	Restoration of heart rhythm
Isoquinoline	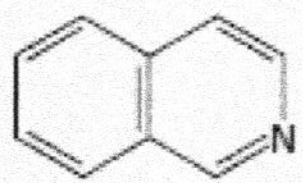	Tyrosine	Codeine	Analgesic (pain relief), treatment of coughs
			Morphine	Analgesic
Indole	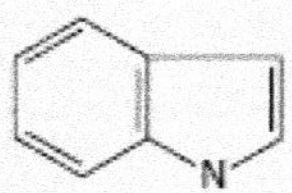	Tryptophan	Psilocybin	Halucinogen
			Reserpine	Treatment of hypertension, treatment of psychoses
			Strychnine	Rat poison, treatment of eye disorders

Figure 7.6: Major Types of Alkaloids, their Amino Acid Precursors, and well-known Examples of each Type.

7.6.2. Cyanogenic Glycosides

Various nitrogenous protective compounds other than alkaloids are found in plants. Two groups of these substances– cyanogenic glycosides and glucosinolates– are not in themselves toxic but are readily broken down to give off volatile poisons when the plant is crushed. Cyanogenic glycosides release the well-known poisonous gas hydrogen cyanide (HCN). Cyanogenic glycosides are not normally broken down in the intact plant because the glycoside and the degradative enzymes are spatially separated, in different cellular compartments or in different tissues. Under ordinary conditions this compartmentation prevents decomposition of the glycoside. When the leaf is damaged, however, as during herbivore feeding, the cell contents of different tissues mix and HCN forms. Cyanogenic glycosides are widely distributed in the plant kingdom and are frequently encountered in legumes, grasses, and species of the rose family.

7.6.3. Glucosinolates

A second class of plant glycosides, called the glucosinolates, or mustard oil glycosides, break down to release volatile defensive substances. Found principally in the Brassicaceae and related plant families, glucosinolates give off the compounds responsible for the smell and taste of vegetables such as cabbage, broccoli, and radishes. The release of these mustard-smelling volatiles from glucosinolates is catalyzed by a hydrolytic enzyme, called a thioglucosidase or myrosinase, that cleaves glucose from its bond with the sulfur atom. The resulting aglycone, the nonsugar portion of the molecule, rearranges with loss of the sulfate to give pungent and chemically reactive products, including isothiocyanates and nitriles, depending on the conditions of hydrolysis. These products function in defense as herbivore toxins and feeding repellents.

7.6.4. Jasmonic Acid

Jasmonic acid is a plant stress hormone that activates many defense in plants, jasmonic acid is synthesized from linolenic acid (18:3), which is released from membrane lipids and then converted to jasmonic acid Responses. Jasmonic acid is known to induce the transcription of a host of genes involved in plant defense metabolism. For example, recent research on the Madagascar periwinkle (Catharanthus roseus), which makes some valuable anticancer alkaloids, identified a transcription factor that responds to jasmonic acid by activating the expression of several genes encoding alkaloid biosynthetic genes (van der Fits and Memelink 2000). Interestingly, this transcription factor also activates the genes of certain primary metabolic pathways that provide precursors for alkaloid formation.

Proteinase inhibitors and certain other defenses are not continuously present in plants, but are synthesized only after initial herbivore or pathogen attack. In tomatoes, insect feeding leads to the rapid accumulation of proteinase inhibitors throughout the plant, even in undamaged areas far from the initial feeding site. The systemic production of proteinase inhibitors in young tomato plants is triggered by a complex sequence of events:

1. Wounded tomato leaves synthesize prosystemin, a large (200 amino acid) precursor protein.
2. Prosystemin is proteolytically processed to produce the short (18 amino acid) polypeptide called systemin, the first (and so far only) polypeptide hormone discovered in plants (Pearce *et al.*, 1991).
3. Systemin is released from damaged cells into the apoplast.
4. Systemin is then transported out of the wounded leaf via the phloem.
5. In target cells, systemin is believed to bind to a site on the plasma membrane and initiate the biosynthesis of jasmonic acid, a plant growth regulator that has wide-ranging effects.
6. Jasmonic acid eventually activates the expression of genes that encode proteinase inhibitors. Other signals, such as ABA (abscisic acid), salicylic acid, and pectin fragments from damaged plant cell walls also appear to participate in this woundsignaling cascade.

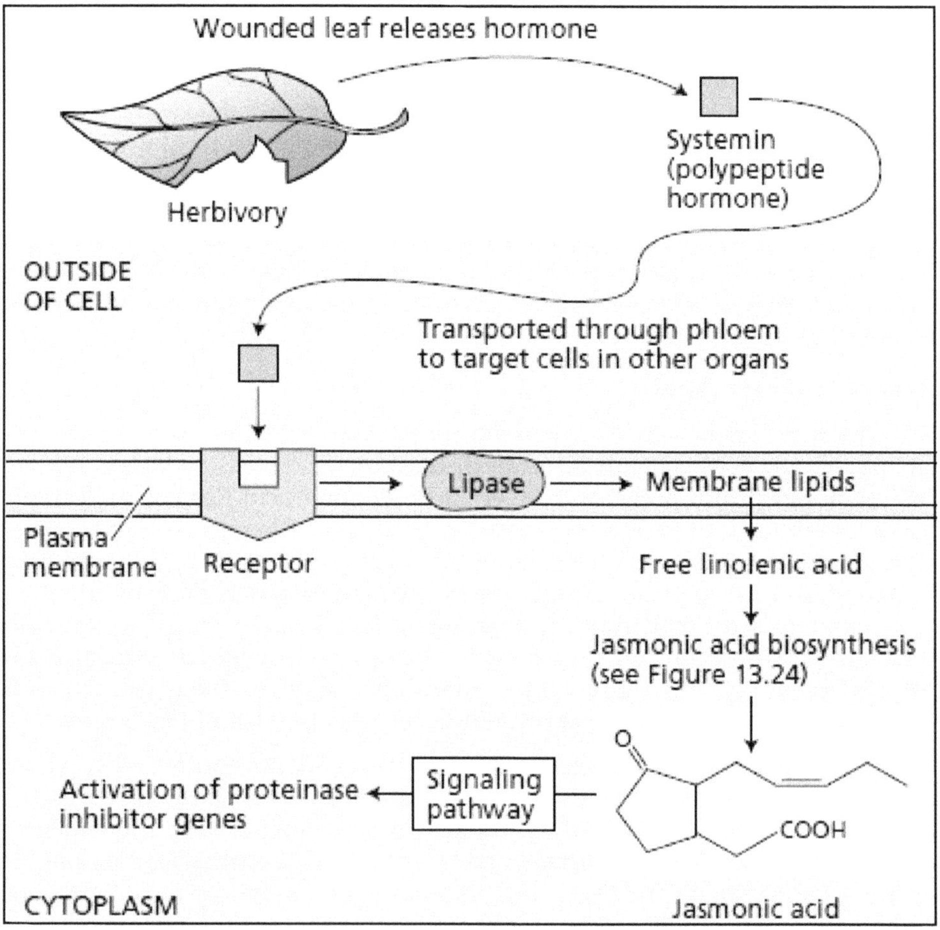

Figure 7.7: Signaling Pathway of Jasmonic Acid (Courtesy of Taiz and Zieger).

7.6.5. Nonprotein Amino Acids

Plants and animals incorporate the same 20 amino acids into their proteins. However, many plants also contain unusual amino acids, called nonprotein amino acids, that are not incorporated into proteins but are present instead in the free form and act as protective substances. Nonprotein amino acids are often very similar to common protein amino acids. Canavanine, for example, is a close analog of arginine, and azetidine-2-carboxylic acid has a structure very much like that of praline.

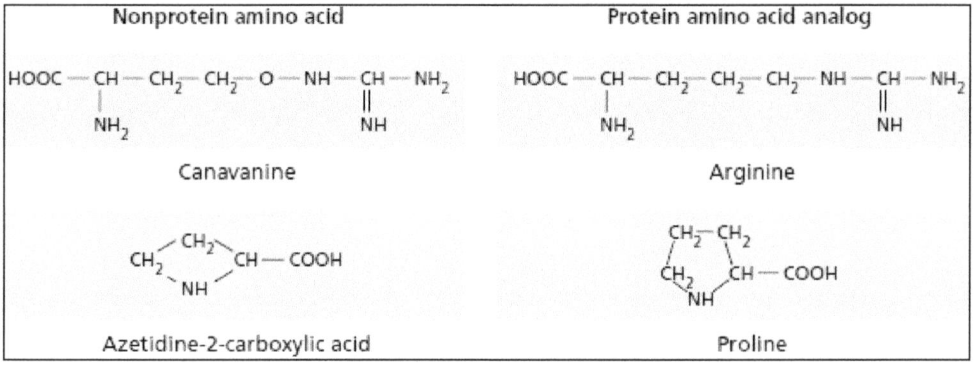

Some block the synthesis or uptake of protein amino acids; others, such as canavanine, can be mistakenly incorporated into proteins. After ingestion, canavanine is recognized by the herbivore enzyme that normally binds arginine to the arginine transfer RNA molecule, so it becomes incorporated into proteins in place of arginine. The usual result is a nonfunctional protein because either its tertiary structure or its catalytic site is disrupted. Plants that synthesize nonprotein amino acids are not susceptible to the toxicity of these compounds. The jack bean (*Canavalia ensiformis*), which synthesizes large amounts of canavanine in its seeds.

References

Cordell, G. (ed.). (1997). *The Alkaloids*. San Diego: Academic Press.

Davies, P.J. (2004). *Plant Hormones: Biosynthesis, Signal Transduction, Action.* Dordrecht: Kluwer Academic Publication.

Herrmann, K.M. and Weaver, L.M. (1999). The shikimate pathway. *Annual Review Plant Physiology, Plant Molecular Biology*, 50: 473–503.

Hartmann, T. (1999). Chemical ecology of pyrrolizidine alkaloids. *Planta*, 207: 483–495.

Chapter 8

Sensory Photobiology

8.1. Control of Flowering

The events occurring in the shoot apex that specifically commit the apical meristem to produce *flowers* are collectively referred to as floral evocation. The developmental signals that bring about floral evocation include endogenous factors, such as *circadian rhythms*, *phase change*, and *hormones*, and external factors, such as day length (*photoperiod*) and temperature (*vernalization*). In the case of photoperiodism, transmissible signals from the leaves, collectively referred to as the floral stimulus, are translocated to the shoot apical meristem. The interactions of these endogenous and external factors enable plants to synchronize their reproductive development with the environment.

8.2. Floral Meristems and Floral Organ Development

Floral meristems usually can be distinguished from vegetative meristems, even in the early stages of reproductive development, by their larger size. The transition from vegetative to reproductive development is marked by an increase in the frequency of cell divisions within the central zone of the shoot apical meristem. In the vegetative meristem, the cells of the central zone complete their division cycles slowly. As reproductive development commences, the increase in the size of the meristem is largely a result of the increased division rate of these central cells. Recently, genetic and molecular studies have identified a network of genes that control floral morphogenesis in Arabidopsis, snapdragon (Antirrhinum), and other species.

As plants initiate reproductive development, the vegetative meristem is transformed into an indeterminate primary inflorescence meristem that produces floral meristems on its flanks. The lateral buds of the cauline leaves (inflorescence leaves) develop into secondary inflorescence meristems, and their activity repeats the pattern of development of the primary inflorescence meristem.

Floral meristems initiate four different types of floral organs: sepals, petals, stamens, and carpels (Coen and Carpenter, 1993). These sets of organs are initiated in concentric rings, called whorls, around the flanks of the meristem (Figure 8.1). The initiation of the innermost organs, the carpels, consumes all of the meristematic cells in the apical dome, and only the floral organ primordia are present as the floral bud develops. In the wild-type Arabidopsis flower, the whorls are arranged as follows:

- ✰ The first (outermost) whorl consists of four sepals, which are green at maturity. The second whorl is composed of four petals, which are white at maturity.
- ✰ The third whorl contains six stamens, two of which are shorter than the other four.
- ✰ The fourth whorl is a single complex organ, the gynoecium or pistil, which is composed of an ovary with two fused carpels, each containing numerous ovules, and a short style capped with a stigma.

Mutations have identified three classes of genes that regulate floral development: floral organ identity genes, cadastral genes, and meristem identity genes. Meristem identity genes must be active for the primordial formed at the flanks of the apical meristem to become floral meristems. (Recall that an apical meristem that is forming floral meristems on its flanks is known as an inflorescence meristem.) For example, mutants of Antirrhinum (snapdragon) that have a defect in the meristem identity gene FLORICAULA develop an inflorescence that does not produce flowers. Instead of causing floral meristems to form in the axils of the bracts, the mutant floricaula gene results in the development of additional inflorescence meristems at the bract axils. The wild-type floricaula (FLO) gene controls the determination step in which floral meristem identity is established.

The floral organ identity genes were identified through homeotic mutations that altered floral organ identity so that some of the floral organs appeared in the wrong place. For example, *Arabidopsis* plants with mutations in the *APETALA2* (*AP2*) gene produce flowers with carpels where sepals should be, and stamens where petals normally appear. The homeotic genes that have been cloned so far encode transcription factors–proteins that control the expression of other genes. Most plant homeotic genes belong to a class of related sequences known as MADS box genes, whereas animal homeotic genes contain sequences called homeobox.

Many of the genes that determine floral organ identity are MADS box genes, including the *DEFICIENS* gene of snapdragon and the *AGAMOUS, PISTILLATA1,* and *APETALA3* genes of *Arabidopsis*. The MADS box genes share a characteristic, conserved nucleotide sequence known as a *MADS box,* which encodes a protein structure known as the *MADS domain*. Not all genes containing the MADS box domain are homeotic genes. For example, AGL20 is a MADS box gene, but it functions as a meristem identity gene. Five different genes are known to specify floral organ identity in *Arabidopsis*: *APETALA1* (*AP1*), *APETALA2* (*AP2*), *APETALA3* (*AP3*), *PISTILLATA* (*PI*), and *AGAMOUS* (*AG*) (Bowman *et al.*, 1989; Weigel and Meyerowitz 1994). These homeotic genes fall into three classes–types A, B, and C–defining three different kinds of activities.

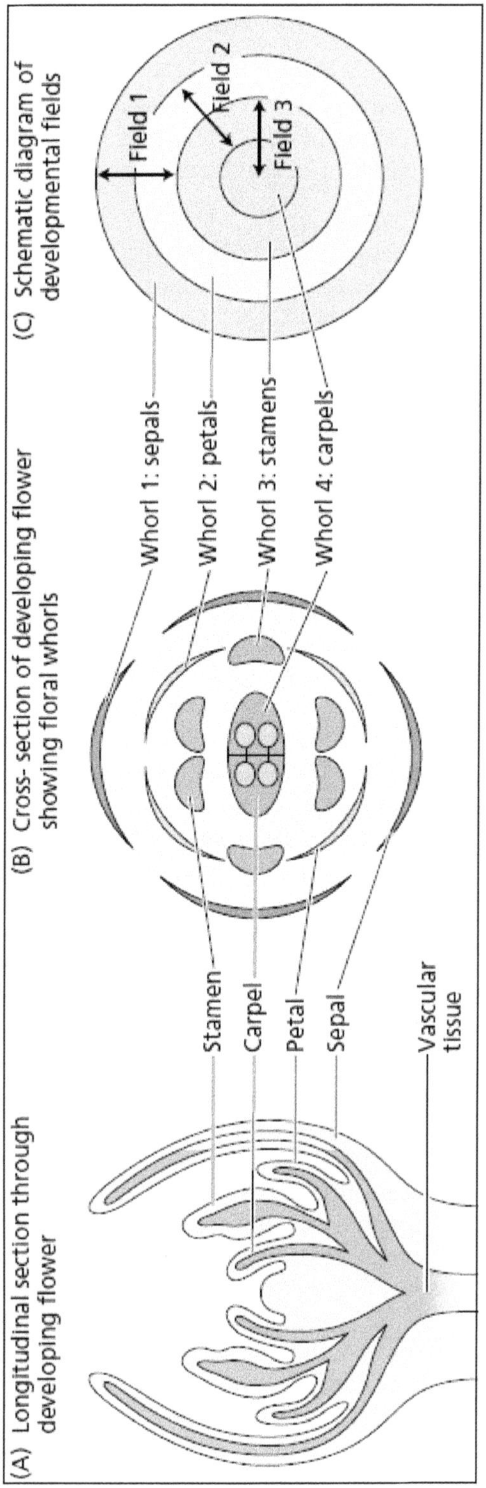

Figure 8.1: Floral Organ are Initiated Sequentially by the Floral Meristem of Arabidopsis.

8.3. The ABC Model Explains the Determination of Floral Organ Identity

In 1991 the ABC model was proposed to explain how homeotic genes control organ identity. The ABC model postulates that organ identity in each whorl is determined by a unique combination of the three organ identity gene activities (Figure 8.2):

- ✰ Activity of type A alone specifies sepals.
- ✰ Activities of both A and B are required for the formation of petals.
- ✰ Activities of B and C form stamens.
- ✰ Activity of C alone specifies carpels.

The model further proposes that activities Aand C mutually repress each other (Figure 8.1); that is, both A- and C-type genes have cadastral function in addition to their function in determining organ identity. The patterns of organ formation in the wild type and most of the mutant phenotypes are predicted and explained by this model. The challenge now is to understand how the expression pattern of these organ identity genes is controlled by cadastral genes; how organ identity genes, which encode transcription factors, alter the pattern of other genes expressed in the developing organ; and finally how this altered pattern of gene expression results in the development of a specific floral organ.

8.4. Flora Evocation

A plant may flower within a few weeks after germinating, as in annual plants such as groundsel (*Senecio vulgaris*). Alternatively, some perennial plants, such as many forest trees, may grow for 20 or more years before they begin to produce flowers. Different species flower at widely different ages, indicating that the age, or perhaps the size, of the plant is an *internal* factor controlling the switch to reproductive development. The case in which flowering occurs strictly in response to internal developmental factors and does not depend on any particular environmental conditions is referred to as *autonomous regulation*. In contrast to plants that flower entirely through an autonomous pathway, some plants exhibit an absolute requirement for the proper environmental cues in order to flower. This condition is termed an *obligate* or *qualitative* response to an environmental cue. In other plant species, flowering is promoted by certain environmental cues but will eventually occur in the absence of such cues. This is called a *facultative* or *quantitative* response to an environmental cue. The flowering of this latter group of plants, which includes *Arabidopsis*, thus relies on both environmental and autonomous flowering systems. Photoperiodism and vernalization are two of the most important mechanisms underlying seasonal responses. *Photoperiodism* is a response to the length of day; vernalization is the promotion of flowering–at subsequent higher temperatures–brought about by exposure to cold. Other signals, such as total light radiation and water availability, can also be important external cues. The evolution of both internal (autonomous) and external (environment-sensing) control systems enables plants to carefully regulate flowering at the optimal time for reproductive success.

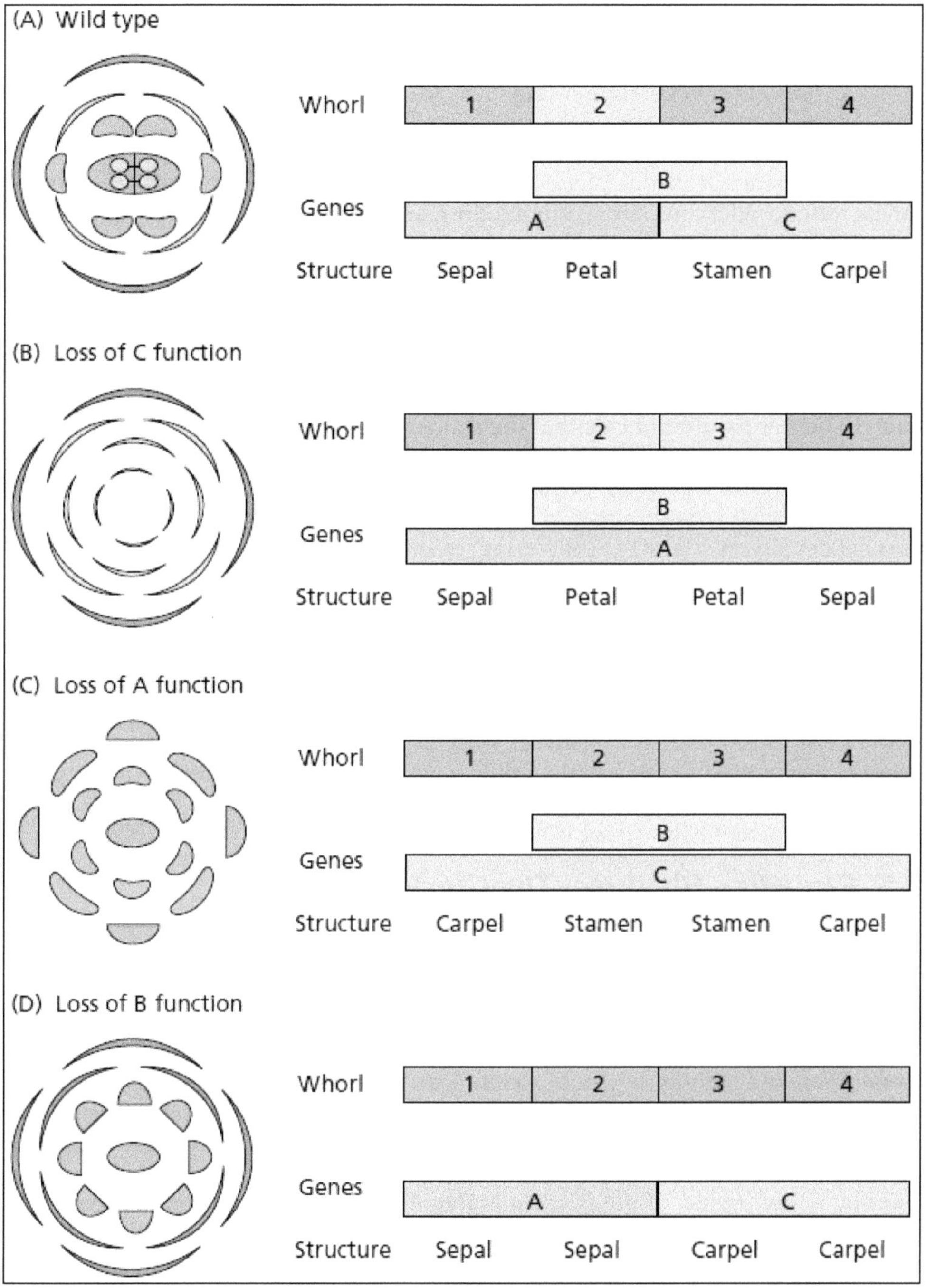

Figure 8.2: ABC Model (Courtsy of Taiz and Zieger).

During postembryonic development, the shoot apical meristem passes through three more or less well-defined developmental stages in sequence:

1. The juvenile phase
2. The adult vegetative phase
3. The adult reproductive phase

The transition from one phase to another is called phase change. Phase changes can be influenced by nutrients, gibberellins, and other chemical signals. The floral evocation requires the apical bud to pass through two developmental stages (Figure 8.3) (McDaniel *et al.*, 1992). One stage is the acquisition of competence. Abud is said to be competent if it is able to flower when given the appropriate developmental signal. For example, if a vegetative shoot (scion) is grafted onto a flowering stock and the scion flowers immediately, it is demonstrably capable of responding to the level of floral stimulus present in the stock and is therefore competent. Failure of the scion to flower would indicate that the shoot apical meristem has not yet attained competence. Thus the juvenile meristems of herbaceous plants are competent to flower, but those of woody species are not. The next stage that a competent vegetative bud goes through is determination. A bud is said to be determined if it progresses to the next developmental stage (flowering) even after being removed from its normal context. Thus a florally determined bud will produce flowers even if it is grafted onto a vegetative plant that is not producing any floral stimulus.

In some cases the expression of flowering may be delayed or arrested even after the apex becomes determined, unless it receives a second developmental signal that stimulates expression (see above figure). For example, intact *Lolium temulentum* (darnel rye grass) plants. In general, once a meristem has become competent, it exhibits an increasing tendency to flower with age (leaf number). For example, in plants controlled by day length, the number of short-day or long-day cycles necessary to achieve flowering is often fewer in older plants.

8.5. Circadian Rhythms: The Clock Within

Organisms are normally subjected to daily cycles of light and darkness, and both plants and animals often exhibit rhythmic behavior in association with these changes. Examples of such rhythms include leaf and petal movements (day and night positions), stomatal opening and closing, growth and sporulation patterns in fungi (*e.g.*, *Pilobolus* and *Neurospora*), time of day of pupal emergence (the fruit fly *Drosophila*), and activity cycles in rodents, as well as metabolic processes such as photosynthetic capacity and respiration rate.When organisms are transferred from daily light–dark cycles to continuous darkness (or continuous dim light), many of these rhythms continue to be expressed, at least for several days. Under such uniform conditions the period of the rhythm is then close to 24 hours, and consequently the term circadian rhythm is applied. Because they continue in a constant light or dark environment, these circadian rhythms cannot be direct responses to the presence or absence of light but must be based on an internal pacemaker, often called an endogenous oscillator. The endogenous oscillator is coupled to a variety of physiological processes, such as leaf movement or photosynthesis, and it maintains

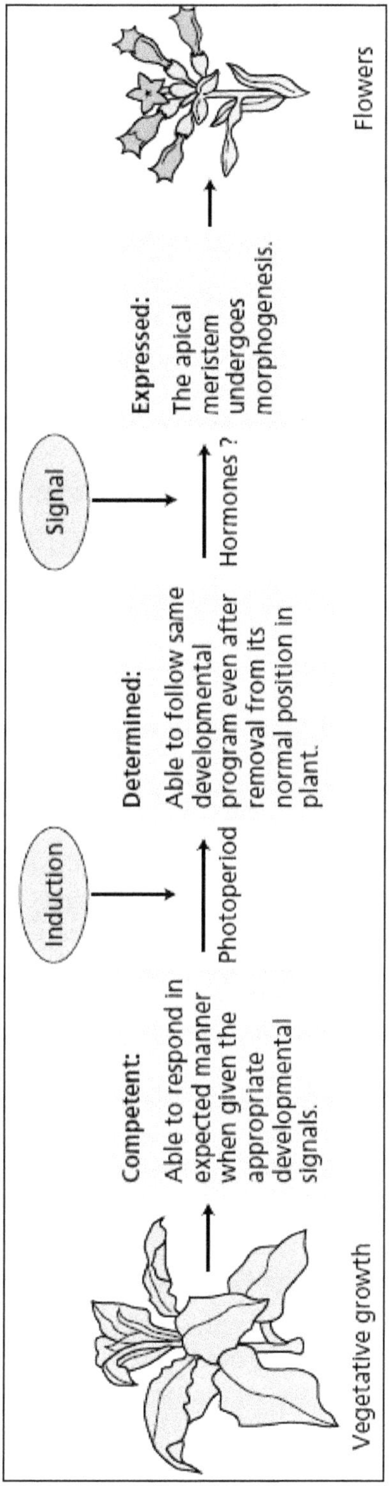

Figure 8.3: A Simple Model for Flower Evocation at the Shoot Apex.

the rhythm. For this reason the endogenous oscillator can be considered the clock mechanism, and the physiological functions that are being regulated, such as leaf movements or photosynthesis, are sometimes referred to as the hands of the clock.

8.6. Characteristic Features of Circadian Rhythms

Circadian rhythms arise from cyclic phenomenons that are defined by three parameters. The amplitude of a biological rhythm can often vary while the period remains unchanged.

1. **Period**, the time between comparable points in the repeating cycle. Typically the period is measured as the time between consecutive maxima (peaks) or minima (troughs) (Figure 8.4).
2. **Phase 2**, any point in the cycle that is recognizable by its relationship to the rest of the cycle. The most obvious phase points are the peak and trough positions.
3. **Amplitude**, usually considered to be the distance between peak

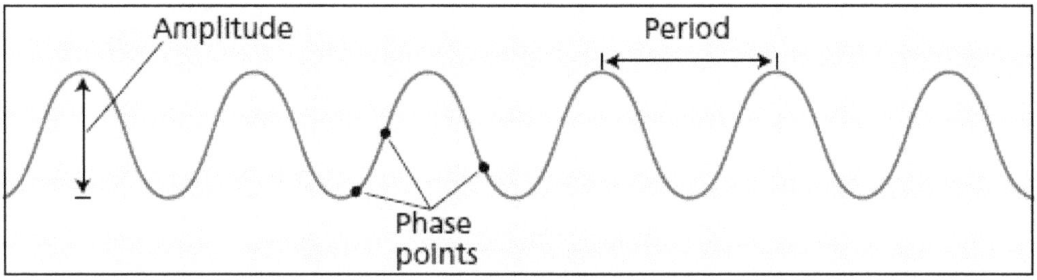

Figure 8.4: A Typical Circadian Rhythms.

In constant light or darkness, rhythms depart from an exact 24-hour period. The rhythms then drift in relation to solar time, either gaining or losing time depending on whether the period is shorter or longer than 24 hours. Under natural conditions, the endogenous oscillator is entrained (synchronized) to a true 24-hour period by environmental signals, the most important of which are the light-to-dark transition at dusk and the dark-to-light transition at dawn.

Although the rhythms are generated internally, they normally require an environmental signal, such as exposure to light or a change in temperature, to initiate their expression. Phase-shifting responses enable the rhythm to be entrained to approximately 24-hour cycles with different durations of light and darkness, and they demonstrate that the rhythm will run differently under different natural conditions of day length.

Phytochromes and Cryptochromes Entrain the Clock

The low levels and specific wavelengths of light that can induce phase shifting indicate that the light response must be mediated by specific photoreceptors rather than rates of photosynthesis. For example, the red-light entrainment of rhythmic nyctonastic leaf movements in Samanea, a semitropical leguminous tree, is a low-fluence response mediated by phytochrome. Arabidopsis has five phytochromes, and all but one of them (phytochrome C) have been implicated in clock entrainment. Each phytochrome acts as a specific photoreceptor for red, far-red, or blue light. In addition, the CRY1 and CRY2 proteins participate in blue-light entrainment of the clock, as they do in insects and mammals (Devlin and Kay, 2000). Surprisingly, CRY proteins also appear to be required for normal entrainment by red light. Since these proteins do not absorb red light, this requirement suggests that CRY1 and CRY2 may act as intermediates in phytochrome signaling during entrainment of the clock (Yanovsky and Kay, 2001).

8.7. Photoperiodism

The circadian clock enables organisms to determine the time of day at which a particular molecular or biochemical event occurs. Photoperiodism, or the ability of an organism to detect day length, makes it possible for an event to occur at a particular time of year, thus allowing for a seasonal response. Circadian rhythms and photoperiodism have the common property of responding to cycles of light and darkness.

Precisely at the equator, day length and night length are equal and constant throughout the year. As one moves away from the equator toward the poles, the days become longer in summer and shorter in winter. Not surprisingly, plant species have evolved to detect these seasonal changes in day length, and their specific photoperiodic responses are strongly influenced by the latitude from which they originated. Photoperiodic phenomena are found in both animals and plants. Plant responses controlled by day length are numerous, including the initiation of flowering, asexual reproduction, the formation of storage organs, and the onset of dormancy. Perhaps all plant photoperiodic responses utilize the same photoreceptors, with subsequent specific signal transduction pathways regulating different responses. Because it is clear that monitoring the passage of time is essential to all photoperiodic responses, a timekeeping mechanism must underlie both the time-of-year and the time-of-day responses. The circadian oscillator is thought to provide an endogenous time-measuring mechanism that serves as a reference point for the response to incoming light (or dark) signals from the environment.

Garner and Henry Allard, conducted experiment in the 1920s at, Maryland. They found that a mutant variety of tobacco, Maryland Mammoth, grew profusely to about 5 m in height but failed to flower in the prevailing conditions of summer. However, the plants flowered in the greenhouse during the winter under natural light conditions. Garner and Allard concluded that the length of the day was the determining factor in flowering and were able to confirm this hypothesis in many different species and conditions.

The classification of plants according to their photoperiodic responses is usually based on flowering, even though many other aspects of plants' development may also be affected by day length. The two main photoperiodic response categories are short-day plants and long-day plants:

1. Short-day plants (SDPs) flower only in short days (qualitative SDPs), or their flowering is accelerated by short days (quantitative SDPs).

2. Long-day plants (LDPs) flower only in long days (*qualitative* LDPs), or their flowering is accelerated by long days (*quantitative* LDPs).

The essential distinction between long-day and shortday plants is that flowering in LDPs is promoted only when the day length *exceeds* a certain duration, called the critical day length, in every 24-hour cycle, whereas promotion of flowering in SDPs requires a day length that is *less than* the critical day length. The absolute value of the critical day length varies widely among species. Long-day plants can effectively measure the lengthening days of spring or early summer and delay flowering until the critical day length is reached. Many varieties of wheat (*Triticum aestivum*) behave in this way. SDPs often flower in fall, when the days shorten below the critical day length, as in many varieties of *Chrysanthemum morifolium*. However, day length alone is an ambiguous signal because it cannot distinguish between spring and fall. Plants exhibit several adaptations for avoiding the ambiguity of day length signal. One is the coupling of a temperature requirement to a photoperiodic response. Certain plant species, such as winter wheat, do not respond to photoperiod until after a cold period (vernalization or overwintering) has occurred. Other plants avoid seasonal ambiguity by distinguishing between *shortening* and *lengthening* days. Such "dual–day length plants" fall into two categories:

1. Long-short-day plants (LSDPs) flower only after a sequence of long days followed by short days. LSDPs, such as *Bryophyllum*, *Kalanchoe*, and *Cestrum nocturnum* (night-blooming jasmine), flower in the late summer and fall, when the days are shortening.
2. Short-long-day plants (SLDPs) flower only after a sequence of short days followed by long days.

SLDPs, such as *Trifolium repens* (white clover), *Campanula medium* (Canterbury bells), and *Echeveria harmsii* (echeveria), flower in the early spring in response to lengthening days.

Finally, species that flower under any photoperiodic condition are referred to as *day-neutral plants*. Day-neutral plants (DNPs) are insensitive to day length. Flowering in DNPs is typically under autonomous regulation–that is, internal developmental control.

Under natural conditions, day and night lengths configure a 24-hour cycle of light and darkness. In principle, a plant could perceive a critical day length by measuring the duration of either light or darknes. The flowering of SDPs is determined primarily by the duration of darkness (Figure 8.5 A). It was possible to induce flowering in SDPs with light periods longer than the critical value, provided that these were followed by sufficiently long nights (Figure 8.5 A). Similarly, SDPs did not flower when short days were followed by short nights. More detailed experiments demonstrated that photoperiodic time keeping in SDPs is a matter of measuring the duration of darkness. For example, flowering occurred only when the dark period exceeded 8.5 hours in cocklebur (*Xanthium strumarium*) or 10 hours in soybean (*Glycine max*). The duration of darkness was also shown to be important in LDPs. These plants were found to flower in short days, provided that the accompanying night length was also short; however, a regime of long days followed by long nights was ineffective.

Night Breaks Can Cancel the Effect of the dark period - A feature that underscores the importance of the dark period is that it can be made ineffective by interruption with a short exposure to light, called a night break. In contrast, interrupting a long day with a brief dark period does not cancel the effect of the long day (see Figure 8.5 B). Night-break treatments of only a few minutes are effective in preventing flowering in many SDPs, including Xanthium and Pharbitis, but much longer exposures are often required to promote flowering in LDPs. In addition, the effect of a night break varies greatly according to the time when it is given. For both LDPs and SDPs, a night break was found to be most effective when given near the middle of a dark period of 16 hours. The discovery of the night-break effect, and its time dependence, had several important consequences. It established the central role of the dark period and provided a valuable probe for studying photoperiodic timekeeping. Because only small amounts of light are needed, it became possible to study the action and identity of the photoreceptor without the interfering effects of photosynthesis and other nonphotoperiodic phenomena. This discovery has also led to the development of commercial methods for regulating the time of flowering in horticultural species, such as Kalanchoe, chrysanthemum, and poinsettia (*Euphorbia pulcherrima*).

Measurements of the effect of a night break on flowering can be used to investigate the role of circadian rhythms in photoperiodic time keeping. For example, when soybean plants, which are SDPs, are transferred from an 8-hour light period to an extended 64-hour dark period, the flowering response to night breaks shows a circadian rhythm. Flowering in SDPs is induced only when exposure to light from a night break or from dawn occurs after completion of the light-sensitive phase of the rhythm. In other words, *flowering is induced when the light exposure is coincident with the appropriate phase of the rhythm*.

The photoperiodic stimulus in both LDPs and SDPs is perceived by the leaves. For example, treatment of a single leaf of the SDP *Xanthium* with short photoperiods is sufficient to cause the formation of flowers, even when the rest of the plant is exposed to long days. Thus, in response to photoperiod the leaf transmits a signal that regulates the transition to flowering at the shoot apex. The photoperiod-regulated processes that occur in the leaves resulting in the transmission of a floral stimulus to the shoot apex are referred to collectively as photoperiodic induction.

Once produced, the flowering stimulus appears to be transported to the meristem via the phloem, and it appears to be chemical rather than physical in nature. Treatments that block phloem transport, such as girdling or localized heat-killing, prevent movement of the floral signal. It is possible to measure rates of movement of the flowering stimulus by removing a leaf at different times after induction, and comparing the time it takes for the signal to reach two buds located at different distances from the induced leaf. The rationale for this type of measurement is that a threshold amount of the signaling compound has reached the bud when flowering takes place, despite the removal of the leaf. Studies using this method

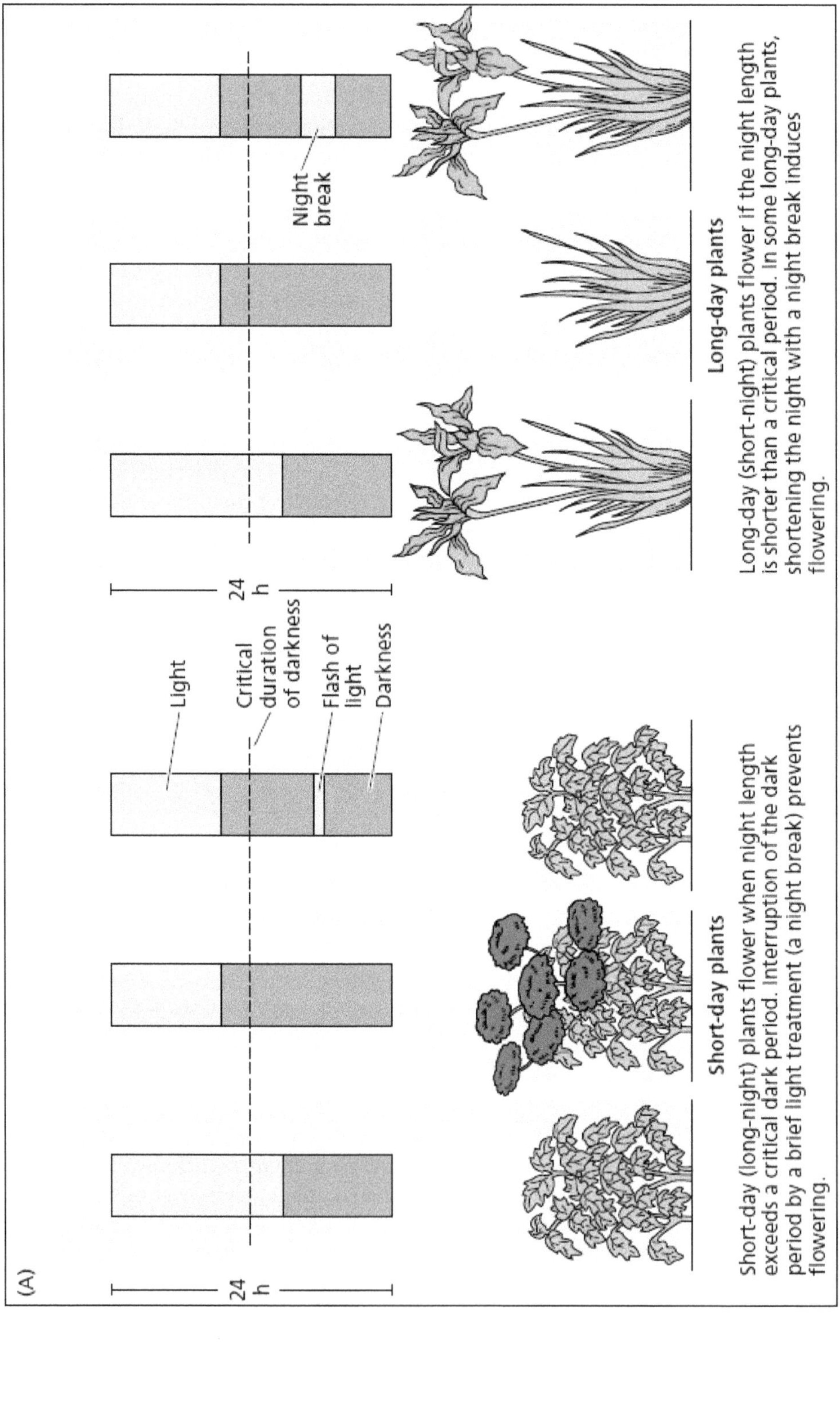

Figure 8.5: Photoperiodic Regulation of Flowering. Effects on SDP and LDP.

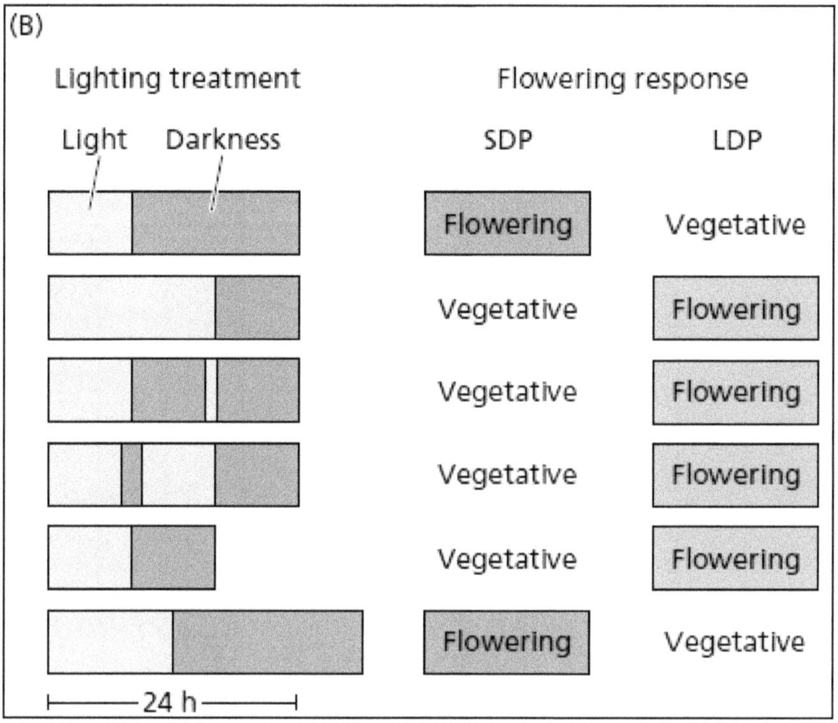

Figure 8.6: Photoperiodic Regulation of Flowering. Effects of the duration of the dark period on the flowering.

have shown that the rate of transport of the flowering signal is comparable to, or somewhat slower than, the rate of translocation of sugars in the phloem

The floral stimulus is translocated along with sugars in the phloem, it is subject to source–sink relations. An induced leaf positioned close to the shoot apex is more likely to cause flowering than an induced leaf at the base of a stem, which normally feeds the roots. Similarly, noninduced leaves positioned between the induced leaf and the apical bud will tend to inhibit flowering by serving as the preferred source leaves for the bud, thus preventing the floral stimulus from the more distal induced leaf from reaching its target.

Role of Phytochrome in the Flower Initiation

In many SDPs, a night break becomes effective only when the supplied dose of light is sufficient to saturate the photoconversion of Pr (phytochrome that absorbs red light) to Pfr (phytochrome that absorbs far-red light). A subsequent exposure to far-red light, which photoconverts the pigment back to the physiologically inactive Pr form, restores the flowering response. In some LDPs, red and far-red reversibility has also been demonstrated. In these plants, a night break of red light promoted flowering, and a subsequent exposure to far-red light prevented this response. flowering is promoted compared with plants maintained under continuous white or red light–a response mediated by PHYB.

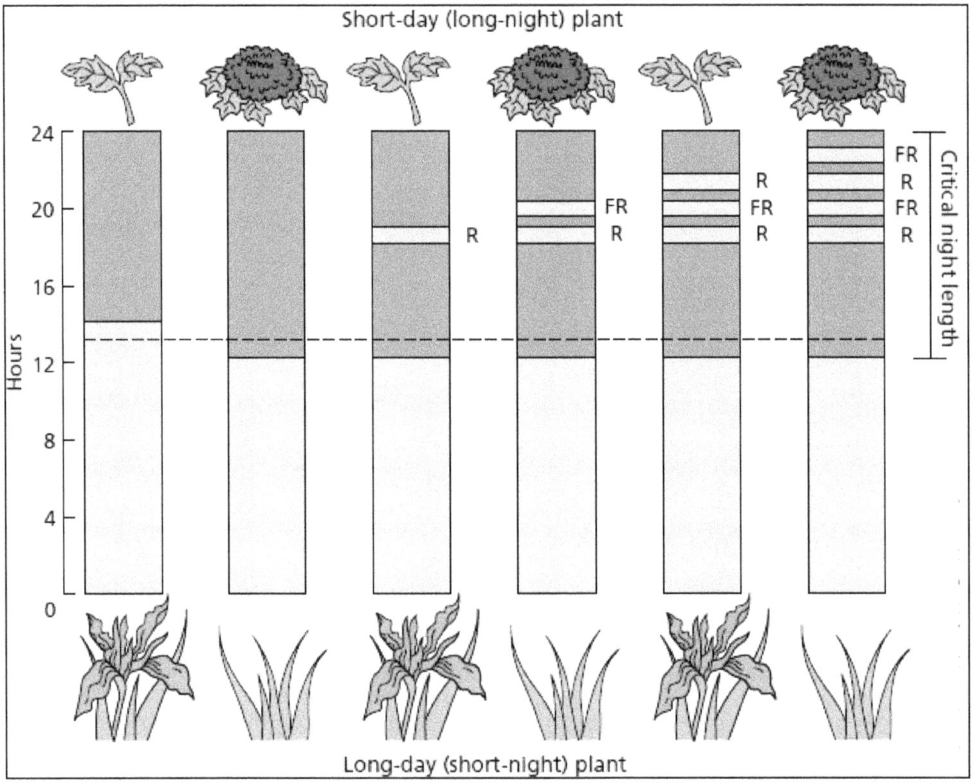

Figure 8.7: Phytochrome Control of Flowering by Red (R) and Far-red (FR) Light.

8.8. The Flowering Stimulus: Florigen Concept

The production in photoperiodically induced leaves of a biochemical signal that is transported to a distant target tissue (the shoot apex) where it stimulates a response (flowering) satisfies an important criterion for a hormonal effect. In the 1930s, Mikhail Chailakhyan, working in Russia, postulated the existence of a universal flowering hormone, which he named florigen. The evidence in support of florigen comes mainly fromearly grafting experiments in which noninduced receptor plants were stimulated to flower by being grafted onto a leaf or shoot from photoperiodically induced donor plants. For example, in the SDP *Perilla crispa*, a member of the mint family, grafting a leaf from a plant grown under inductive short days onto a plant grown under noninductive long days causes the latter to flower. Moreover, the floral stimulus seems to be the same in plants with different photoperiodic requirements. Thus, grafting an induced leaf from the LDP *Nicotiana sylvestris*, grown under long days, onto the SDP Maryland Mammoth tobacco caused the latter to flower under noninductive (long day) conditions. In a few cases, flowering has been induced by grafts between different genera. The SDP Xanthium strumarium flowered under long-day conditions when shoots of flowering Calendula officinalis were grafted onto a vegetative Xanthium stock. There are some following evidences in support of the floral hormone:

- ☆ Photoperiodic stimulus is perceived by leaves.
- ☆ Stimulus may be transmitted from a single leaf.
- ☆ Flowering stimulus can be transmitted from one branch to other branch or cross a graft union.
- ☆ The stimulus is transmitted through the stem to the apical meristem.
- ☆ Transport of stimulus occurs through living tissue.
- ☆ Flowering stimulus is similar both in SDP and LDP

In at least three cases–*Xanthium* (SDP), *Bryophyllum* (SLDP), and *Silene* (LDP)–the induced state appears to be self-propagating (Zeevaart, 1976). That is, young leaves that develop on the receptor plant after it has been induced to flower by a donor leaf can themselves be used as donor leaves in subsequent grafting experiments, even though these leaves have never been subjected to an inductive photoperiod. This phenomenon is called *indirect induction*. It is characteristic of indirect induction that the strength of the floral stimulus from the donor leaf remains constant even after serial grafting of new donors to several plants has taken place.

Exogenous gibberellin can evoke flowering when applied either to rosette LDPs like *Arabidopsis*, or to dual–day length plants such as *Bryophyllum*, when grown under short days. In addition to GAs, other growth hormones can either inhibit or promote flowering. One commercially important example is the striking promotion of flowering in pineapple (*Ananas comosus*) by ethylene and ethylene-releasing compounds–a response that appears to be restricted to members of the pineapple family (Bromeliaceae).

The transition to flowering is regulated by multiple signals and multiple pathways. In *Arabidopsis*, flowering is controlled by four pathways: the photoperiodic, autonomous/vernalization, sucrose, and GA pathways. All of these pathways converge to regulate the meristem identity genes *AGAMOUS-LIKE 20* (*AGL20*) and *LEAFY* (*LFY*). *AGL20* and *LFY*, in turn, regulate the floral homeotic genes to produce the floral organs.

8.9. Vernalization

Many species, especially many biennial and perennial are induced or promoted to flower by low temperature. Lysenko (1938) used term vernalization for a low temperature promotion of flowering in plants. The low temperature requirement for flowering was first noticed by Klippart in 1857 while working with two varities of wheat – the winter wheat and the spring wheat. The winter wheat is sown in winter and produces flower in summers. The spring wheat is sown in the spring and produces flower in summers. He observed that if the germinated seeds of winter wheat are treated with low temperature they can behave as spring wheat. He concluded that winter wheat require low temperature (0-5°C) for subsequent flowering, the phenomenon known as vernalization.

Vernalization is the process whereby flowering is promoted by a cold treatment given to a fully hydrated seed (*i.e.*, a seed that has imbibed water) or to a growing plant. Dry seeds do not respond to the cold treatment. Without the cold treatment,

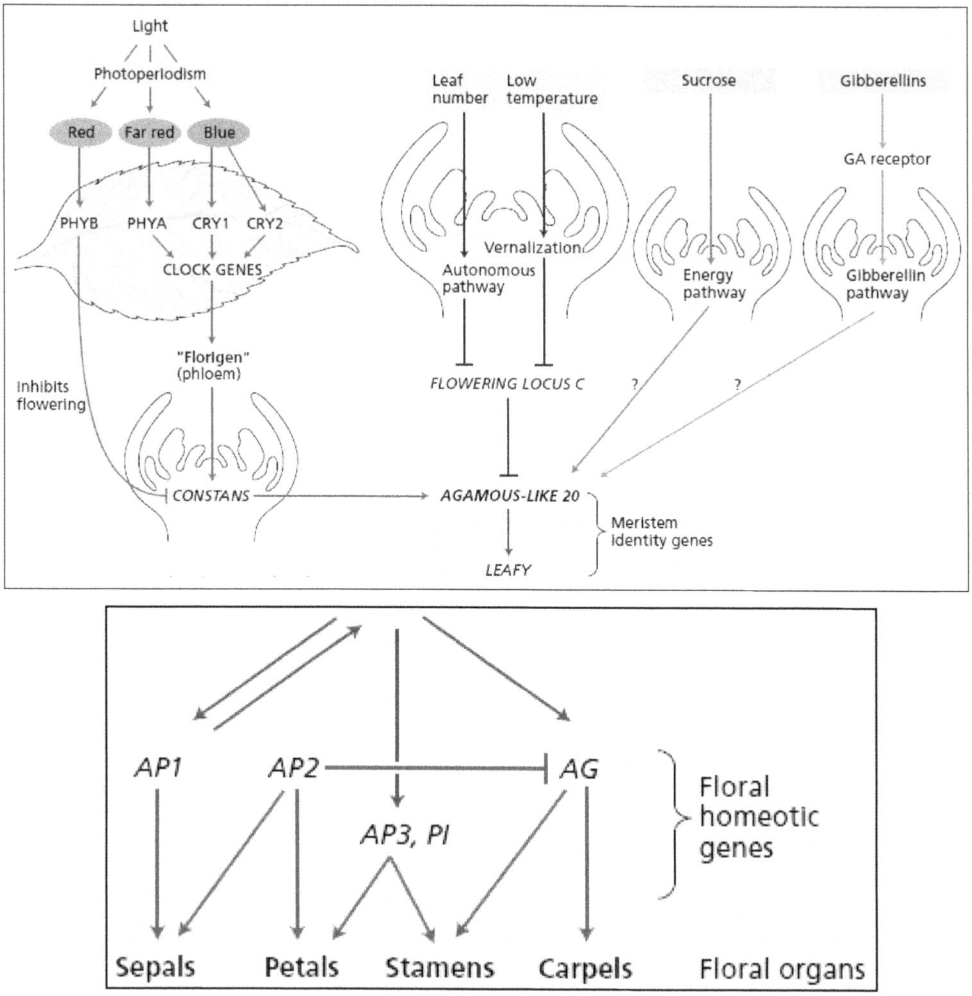

Figure 8.8: Four Developmental Pathway for Flowering in Arabidopsis (Courtesy of Taiz and Zieger).

plants that require vernalization show delayed flowering or remain vegetative. In many cases these plants grow as rosettes with no elongation of the stem.

Winter annuals, such as the winter forms of cereals (which are sown in the fall and flower in the following summer), respond to low temperature very early in their life cycle. They can be vernalized before germination if the seeds have imbibed water and become metabolically active. Other plants, including most biennials (which grow as rosettes during the first season after sowing and flower in the following summer), must reach a minimal size before they become sensitive to low temperature for vernalization. The effective temperature range for vernalization is from just below freezing to about 10°C, with a broad optimum usually between about 1 and 7°C (Lang, 1965). The effect of cold increases with the duration of the cold treatment until the response is saturated. The response usually requires several weeks of exposure to low temperature, but the precise duration varies widely with species and variety. Vernalization can be lost as a result of exposure to devernalizing conditions, such as high temperature, but the longer the exposure to low temperature, the more permanent the vernalization effect. Vernalization appears to take place primarily in the shoot apical meristem. Localized cooling causes flowering when only the stem apex is chilled, and this effect appears to be largely independent of the temperature experienced by the rest of the plant. Excised shoot tips have been successfully vernalized, and where seed vernalization is possible, fragments of embryos consisting essentially of the shoot tip are sensitive to low temperature.

The botanist G. Melchers demonstrated that the product of vernalization could be transmitted from a vernalized to an unvernalized Hyoscymus plant through a graft union. He named this substance as vernal. Later A. Lang *et al.* demonstrated that application of gibberellins can replace the cold treatment for vernalization in many plants.

For vernalization to occur, active metabolism is required during the cold treatment. Sources of energy (sugars) and oxygen are required, and temperatures below freezing at which metabolic activity is suppressed are not effective for vernalization. Furthermore, cell division and DNA replication also appear to be required. One model for how vernalization affects competence is that there are stable changes in the pattern of gene expression in the meristem after cold treatment. Changes in gene expression that are stable even after the signal that induced the change (in this case cold) is removed are known as epigenetic regulation. Epigenetic changes of gene expression in many organisms, from yeast to mammals, often require cell division and DNA replication, as is the case for vernalization. The involvement of epigenetic regulation in the vernalization process has been confirmed in the LDP *Arabidopsis.*

8.10. Phytochrome

Phytochrome, a blue protein pigment with a molecular mass of about 125 kDa (kilodaltons). Phytochrome can interconvert between Pr and Pfr Forms - In dark-grown or etiolated plants, phytochrome is present in a red light–absorbing form, referred to as Pr because it is synthesized in this form. Pr, is converted by red light to a far-red light–absorbing form called Pfr, which is blue-green. Pfr, in turn, can be converted back to Pr by far-red light. Known as photoreversibility, this conversion/ reconversion property is the most distinctive property of phytochrome.

The interconversion of the Pr and Pfr forms can be measured in vivo or in vitro. In fact, most of the spectral properties of carefully purified phytochrome measured in vitro are the same as those observed in vivo. When Pr molecules are exposed to red light, most of them absorb it and are converted to Pfr, but some of the Pfr also absorbs the red light and is converted back to Pr because both Pr and Pfr absorb red light (Figure - 8.9). Thus the proportion of phytochrome in the Pfr form after saturating irradiation by red light is only about 85 per cent. Similarly, the very small amount of far-red light absorbed by Pr makes it impossible to convert Pfr entirely to Pr by broadspectrum far-red light. Instead, an equilibrium of 97 per cent Pr and 3 per cent Pfr is achieved. This equilibrium is termed the photostationary state. In addition to absorbing red light, both forms of phytochrome absorb light in the blue region of the spectrum.

Pfr is the Physiologically Active form of Phytochrome - Because phytochrome responses are induced by red light, they could in theory result either from the appearance of Pfr or from the disappearance of Pr. Evidence such as this has led to the conclusion that Pfr is the physiologically active form of phytochrome. In cases in which it has been shown that a phytochrome response is not quantitatively related to the absolute amount of Pfr, it has been proposed that the ratio between Pfr and Pr, or between Pfr and the total amount of phytochrome, determines the magnitude of the response. The conclusion that Pfr is the physiologically active form of phytochrome.

Phytochrome is a Dimer Composed of Two Polypeptides

Native phytochrome is a soluble protein with a molecular mass of about 250 kDa. It occurs as a dimer made up of two equivalent subunits. Each subunit consists of two components: a light-absorbing pigment molecule called the chromophore, and a polypeptide chain called the apoprotein. The apoprotein monomer has a molecular mass of about 125 kDa. Together, the apoprotein and its chromophore make up the holoprotein. In higher plants the chromophore of phytochrome is a linear tetrapyrrole termed phytochromobilin. There is only one chromophore per monomer of apoprotein, and it is attached to the protein through a thioether linkage to a cysteine residue (Figure 8.9). The polypeptide folds into two major domains separated by a "hinge" region. The larger N-terminal domain is approximately 70 kDa and bears the chromophore; the smaller C-terminal domain is approximately 55 kDa and contains the site where the two monomers associate with each other to form the dimer.

The phytochrome apoprotein alone cannot absorb red or far-red light. Light can be absorbed only when the polypeptide is covalently linked with phytochromobilin to form the holoprotein. Phytochromobilin is synthesized inside plastids and is derived from 5-aminolevulinic acid via a pathway that branches from the chlorophyll biosynthetic pathway. It is thought to leak out of the plastid into the cytosol by a passive process.Assembly of the phytochrome apoprotein with its chromophore is autocatalytic; that is, it occurs spontaneously.

Because the chromophore absorbs the light, conformational changes in the protein are initiated by changes in the chromophore. Upon absorption of light, the Pr chromophore undergoes a *cis–trans* isomerization of the double bond between carbons 15 and 16 and rotation of the C14–C15 single bond (see Figure 8.9) (Andel *et al.*, 1997). During the conversion of Pr to Pfr, the protein moiety of the phytochrome holoprotein also undergoes a subtle conformational change. Several lines of evidence suggest that the light-induced change in the conformation of the polypeptide occurs both in the N-terminal chromophore-binding domain and in the C-terminal region of the protein.

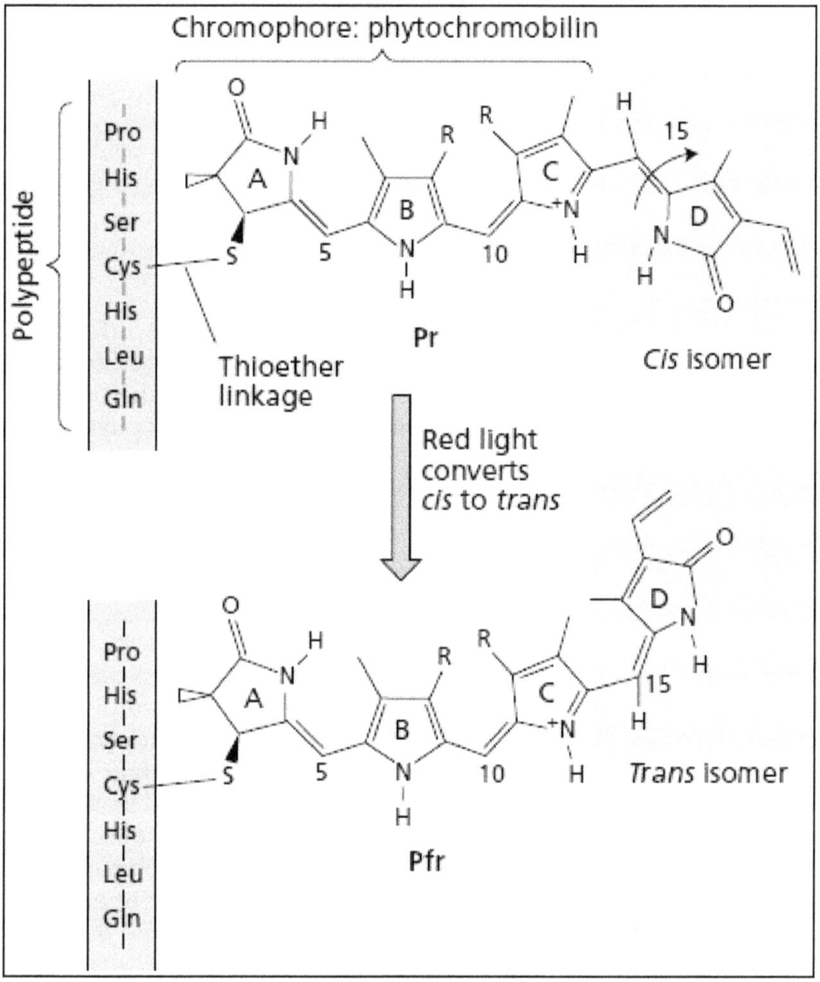

Figure 8.9: Structure of Pr and Pfr Form of the Chromophore (Phytochromobilin) and the Peptide Region Bound to the Chromophore through a Thioester Linkage.

There are two different classes of phytochrome with distinct properties. These have been termed Type I and Type II phytochromes (Furuya 1993). Type I is about nine times more abundant than Type II in dark-grown pea seedlings; in light-grown pea seedlings the amounts of the two types are about equal. More recently, the two types have been shown to be distinct proteins. Phytochrome is encoded by multigene family. phytochrome gene family is named *PHY*, and its five individual members are *PHYA, PHYB, PHYC, PHYD,* and *PHYE.* The apoprotein by itself (without the chromophore) is designated PHY; the holoprotein (with the chromophore) is designated phy. *PHYA* is the only gene that encodes a Type I phytochrome. The Pfr form of the protein encoded by the *PHYA* gene, called PfrA, is unstable.

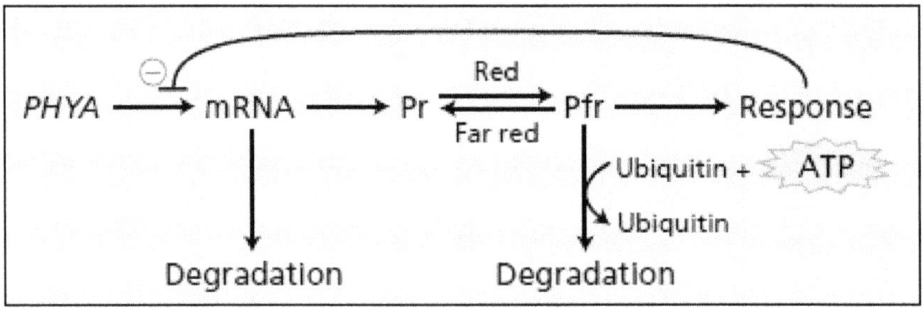

The remaining *PHY* genes (*PHYB* through *PHYE*) encode the Type II phytochromes. Although detected in green plants, these phytochromes are also present in etiolated plants. The reason is that the expression of their mRNAs is not significantly changed by light, and the encoded phyB through phyE proteins are more stable in the Pfr form than is PfrA.

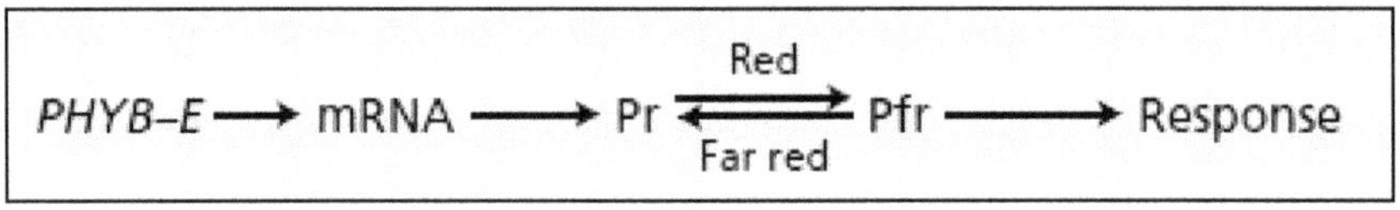

The majority of phytochrome regulated genes are nuclear genes encoding for mRNA s of chloroplast proteins., *e.g.* the small subunit of ribulose1,5-bisphosphate carboxylase/oxygenase (Rubisco) and the light –harvesting chlorophyll a/b binding proteins (LHCP). Rubisco is multimeric protein, consisting of eight identical large sub units (LSU) and eight small subunit (SSU). The LSU encoded by the chloroplast genome and synthesized in the chloroplast. The SSU is encoded in nuclear genes and synthesized in cytoplasm as a precursor polypeptide. The polypeptide of LHCP apoprotein (also known as chlorophyll a/b binding protein) are encoded in the nucleus, synthesized on cytoplasmic ribosome and imported into the chloroplast.

In addition to being distinguished by lag times and escape times, phytochrome responses can be distinguished by the amount of light required to induce them. The amount of light referred to as the influence, which is defined as the number of photons impinging on a unit surface area. The most commonly used units for fluence are moles of quanta per square meter (mol m^{-2}). In addition to the fluence, some phytochrome responses are sensitive to the irradiance, 2 or fluence rate, of light. The units of irradiance in terms of photons are moles of quanta per square meter per second (mol m^{-2} s^{-1}). Each phytochrome response has a characteristic range of light fluences over which the magnitude of the response is proportional to the fluence. These responses fall into three major categories based on the amount of light required: very-low-fluence responses (VLFRs), low-fluence responses (LFRs), and high-irradiance responses (HIRs).

8.11. Dormancy of Seeds

The seeds of most of the plants germinate when they are provided with favourable environmental conditions (such as moisture, air, suitable temperature

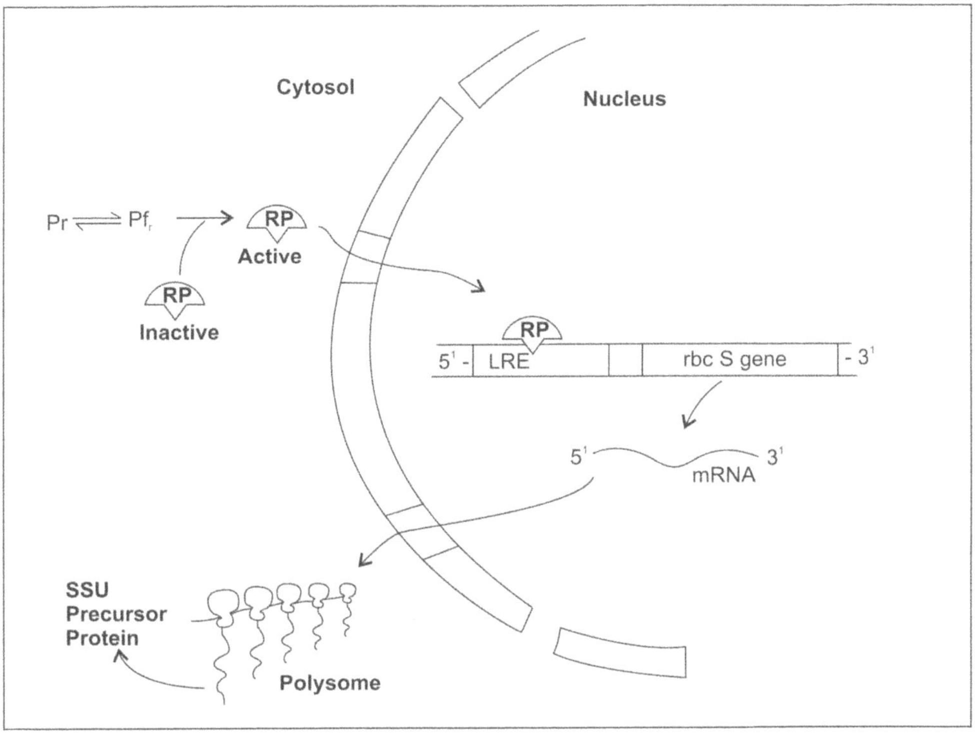

Figure 8.10: Phytochrome Regulation of Rubisco Small Subunit (Courtesy of Hopkins).

and proper sequence of light and dark etc.). There are, however, many more plants whose seeds, although viable, do not germinate even though they get all the favourable environmental conditions.Germination of such seeds may be delayed for days, weeks, month or even years. The seeds of such plants are said to be in a state of dormancy. The dormancy may thus be defined as the condition of seed when it fails to germinate even though the environmental conditions, usually considered favourable for active growth, are present.

Germination of seeds may not always be prevented due to dormancy.Some time seeds fail to germinate because the environmental conditions normally required for growth (*i.e.* sufficient moisture and temperature) are not present.This situation is known as quiescence.

Factors Causing Dorancy of Seeds

1. Dormancy Due to Seed Coats

The seed coat of most of the seeds is composed of several layers of cells.During seed ripening the seed coat becomes dehydrated and forms a hard tough protective covering around the embryo.Because of the undermentioned properties the seeds coat causes barrier for the seeds to germinate –

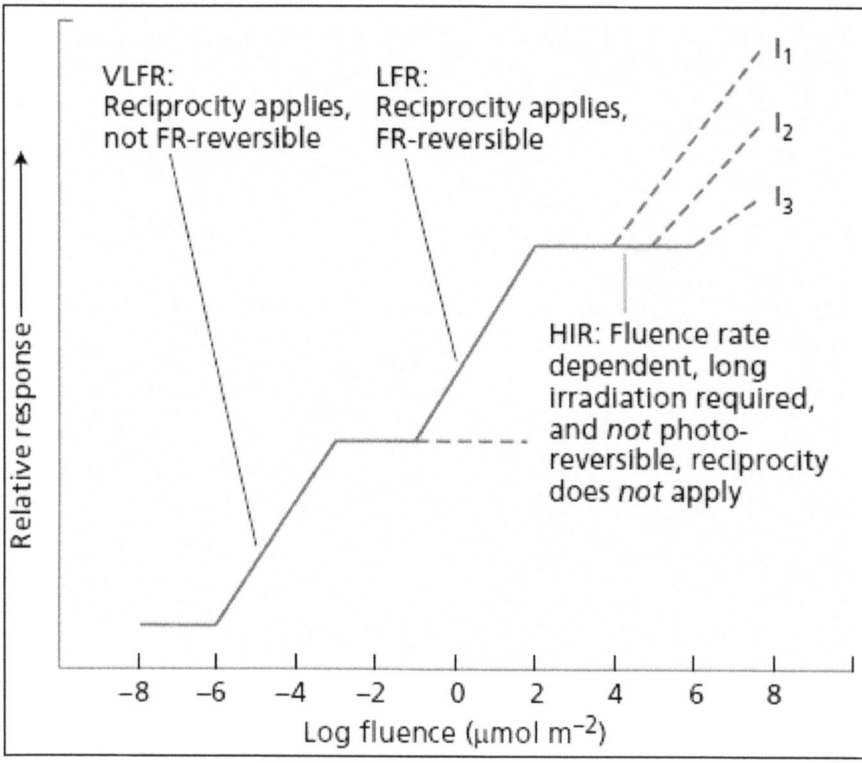

Figure 8.11

- Water impermeability – Seed coat of many plants are impermeable to water. Since the water is primary factor for seed germination,the seeds fail to germinate until the seed coat decays by soil fungi and bacteria (*e.g.* seed of family leguminaceae).
- Gas permeability – The seed coats of some seed are impermeable to gases such as oxygen and carbon dioxide (*e.g.* seeds of Xanthium).
- Mechanical resistance – Some plants have such hard and tough seed coat that it cannot broken by the germinating embryo.Recently, it has been indicated that the mechanical resistance of the seed coats of many seed may be the primary factors for dormancy (*e.g.* seeds of Amaranthus, Capsella etc.

2. Dormancy Due to Condition of Embryo

Seed dormancy due to condition of embryo may be of two kinds. In the first case, the embryo are immature when the seeds ripe *i.e.*, the embryos are rudimentary and poorly developed. Further development of these embryos occurs during the period of dormancy. In the second case, although the embryo are fully developed when seeds mature, but they are unable to germinate promptly. Such seed can be induced to germinate if they are stored in moist, well aerated and low temperature condition. This process is known as stratification or after ripening.

3. Dormancy Due to Specific Light Requirement

The seeds of certain species namely *Lactuca sativa, Nicotiana tabaccum, Lythrum salicaria* etc., have a light requirement for germination. Germination of water imbibed lettuce seeds is stimulated by red light and is inhibited by far-red light.This shows the involvement of photoreversible pigment phytochrome in the process of germination.

4. Dormancy Due to Chilling Requirement

Seeds of many temperate plants, such as peach,plum,cherry etc., require a chilling(low temperature) treatment before they become capable of germinating.The most suitable temperature for chilling treatment is 0–5°C. The chilling requirement in some seeds is very peculiar. They germinate in autumn and develop a young root system but their epicotyls fails to grow.The epicotyls remains dormant till it receves low temperature treatment during winter.This is called epicotyls dormancy. (*e.g.* oak). However, certain seeds require chilling treatment twice,once for breaking rootor radical dormancy and second for epicotyls dormancy (*e.g.* polygonatum).

5. Dormancy Due to Germination Inhibitors

A large number of chemical substances have been isolated from the seeds.Some of them inhibit seed germination,while others promote growth. These substances are organic acids,phenolics, tannins,alkaloids, unsaturated lactones,mustard oil,ammonia-releasing substances,cyanide releasing substances present in the seed covering,block the growth of embryo.

Method of Breaking Seed Dormancy

The dormancy of seeds can be broken and the dormant seed can be induced to germinate by different given methods:

1. Scarification

The dormancy of seeds caused by hard seed coats,can be removed by breaking,softening or weakening of seed coats.The method employed in softening or weakening the seed coat is known as scarification.Under natural condition in the soil,some of the microorganism act upon seed coats and usually decompose them.However various mechanical and chemical treatments have been used to prompt the germination.These are cutting or chipping of seed coat, storage at high temperature,use of organic solvents to remove waxy or fatty compounds.

2. Impaction

In some of seeds water and oxygen are unable to penetrate because entry is blocked by a cork-like filling in the small opening in the seed coat.These seed are shaken vigorously to remove the plug. The treatment is called impaction. It has been applied to seeds of *Crotolaria egyptica, Melilotus alba, Trigonella arabica* etc.

3. Stratification

The dormancy caused due to condition of embryo has been noticed in the seeds of rosaceous plants (peach, apple, plum, cherry etc.). These seeds will not germinate until they have been exposed in well aerated,moist condition under low temperature (0–10°C) for weeks to months. Such a treatment is known as stratification. Some time the process is called as after ripening. It is believed that the embryo f such seed which require stratification are immature in the sense that some chemical changes occure during this period which are required for seed germination.Some of the chemical changes during stratification are:

- ☆ Shifts in the concentration of nitrogen and phosphorus between the various parts of the seeds.
- ☆ Shifting of constituent amino acids,organic acids and enzyme
- ☆ Decomposition of cyanogenic glycosides in rosaceae seeds.
- ☆ Change in the concentration of various growth regulators.

4. Alternating Temperature

5. Light

The seeds which respond to light for their germination are known as photoblastic. There are three categories of such seeds:

- ☆ Single exposure of light promotes seeds germination. (Positive photoblastic seeds).
- ☆ Light inhibits seed germination *i.e.* seeds require complete darkness for germination (negative photoblastic seeds), and
- ☆ Seeds germinate in either light or dark (nonphotoblastic seeds).

8.12. Bud Dormancy

Dormancy of bud is defined as cessation of observable growth. Buds of many herbaceous and woody perennials of temperate regions enter a dormant state in late summer. The bud dormancy of this type is usually broken by cold temperature. The bud dormancy is somewhat different than seeds dormancy.The bud remains attached to the plant and faces adverse environmental condition.It is exposed to natural condition which prevent physiological activity.The bud is well hydrated and not enclosed within a hard covering.

The primary factors for induction of bud dormancy is photoperiodism.The other factors are temperature,nutrition,water status and irradiance.Usually the winter bud dormancy is caused by short day length and relieved by long day lengths.The photoperiodic stimulus is perceived by bud scales and leaves and the pigment phytochrome is involved as mediator of photoperiodic induction.Bud dormancy in woody trees is controlled by growth inhibitory substances,such as,abscisic acid is mainly responsible for causing bud dormancy. However, recent studies indicate that abscisic acid has no effects on the induction of bud dormancy. It probably maintains the period of dormancy. Bud dormancy can be broken by the treatment of ethylene.

8.13. Seed Germination

Structure of Seed

Seed is the miniature of plant body. It consists of an embryo enclosed by protective seed coat.Some seed are provided with food materials in the form of endosperm tissue surrounding the embryo. While others the food materials is stored in cotyledons.

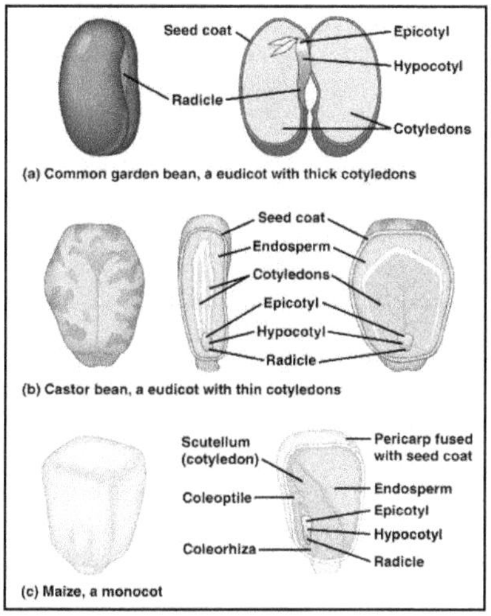

Figure 8.12: Stages of Seed Germination.

The embryo consists of radical and plumule.The radical when elongates gives rise to primary root whereas the plumule elongates to produce aerial shoot. When the embryo contains two cotyledons, the seed is said to be dicot and when it contains only one cotyledon it is called monocotyledonous.

Type of Germination

Germination of seed is of following types – Hypogeal and epigeal etc.

Dicot Germination

The part of the plant that first emerges from the seed is the embryonic root, termed the radicle or primary root. It allows the seedling to become anchored in the ground and start absorbing water. After the root absorbs water, an embryonic shoot emerges from the seed. This shoot comprises three main parts: the cotyledons (seed leaves), the section of shoot below the cotyledons (hypocotyl), and the section of shoot above the cotyledons (epicotyl). The way the shoot emerges differs among plant groups.

Epigeal

In epigeal germination (or epigeous germination), the hypocotyl elongates and forms a hook, pulling rather than pushing the cotyledons and apical meristem through the soil. Once it reaches the surface, it straightens and pulls the cotyledons and shoot tip of the growing seedlings into the air. Beans, tamarind, and papaya are examples of plants that germinate this way.

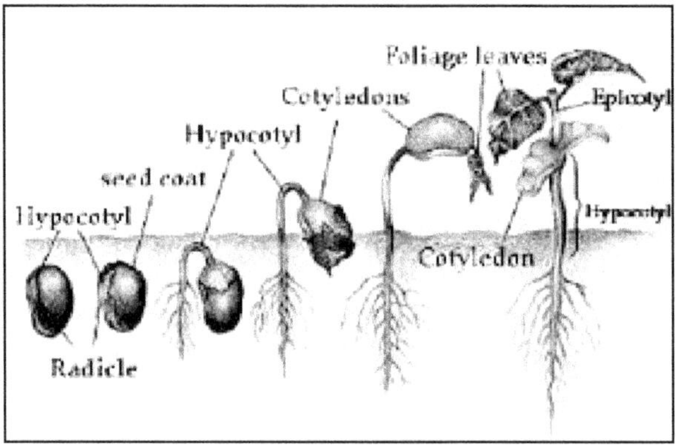

Figure 8.13: Process of Epigeal Germination.

Hypogeal

Germination can also be done by hypogeal germination (or hypogeous germination), where the epicotyl elongates and forms the hook. In this type of germination, the cotyledons stay underground where they eventually decompose. Peas, gram and mango, for example, germinate this way.

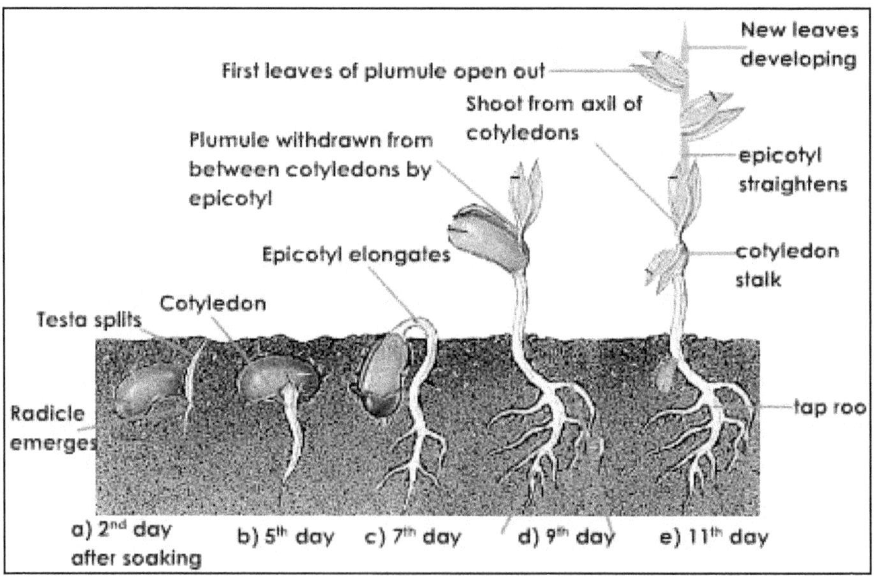

Figure 8.14: Process of Hypogeal Germination.

Monocot Germination

In monocot seeds, the embryo's radicle and cotyledon are covered by a coleorhiza and coleoptile, respectively. The coleorhiza is the first part to grow out of the seed, followed by the radicle. The coleoptile is then pushed up through the ground until it reaches the surface. There, it stops elongating and the first leaves emerge.

Precocious Germination

Actually it is a stage of the growth of the seed, before the germination starts and seed is in the stage of formation. The ripe seeds are highly dehydrated, the water supply is the primary factor needed for germination. The second important factor is temperature.The temperature occure varies from one species to another. Besides, most of the seeds require adequate aeration for germination. Light is not essential factor for the germination of most of the seeds but some light sensitive seeds do require light as a factor.

Physiology of Seed Germination

The initial step that occure in germination is absorption of water,involving imbibitions and osmosis. Nondormant viable seeds,when get conditions favourable for germination imbibe water through the micropylar pore and or the hilum of the seed coat. It results hydration of subcellular organelles leading changes in their organization. Hydration is followed by diverse enzyme activities. The respiratory enzyme, which is already present in quite high level,resume their activity after hydration. Substrate for rapid metabolism,present in the form of storage reserves,are usually located in the endosperm of monocot and cotyledons of dicot seeds.These storage reserves consists of lipids,protein and carbohydrates.The hydrolysis of storage reserves is brought about by the activity of hydrolytic enzymes. As soon as,the insoluble reserves are converted to soluble forms,they are mobilized and translocated to embryonal axis. The control of endosperm activity in germinating seed is hormonal control.The hormoe are usually synthesized and originate in embronal axis which is translocated to storage tissue and causes de-novo synthesis of certain enzyme, *e.g.* gibberellins.

8.14. Sensecence and Programmed Cell Death

Every autumn, people who live in temperate regions can enjoy the beautiful color changes that precede the loss of leaves from deciduous trees. The leaves change color because changing day length and cooling temperatures trigger developmental processes that lead to leaf senescence and death. Senescence is distinct from necrosis, although both senescence and necrosis lead to death. Necrosis is death brought about by physical damage, poisons, or other external injury. In contrast, senescence is a normal, energy-dependent developmental process that is controlled by the plant's own genetic program. Leaves are genetically programmed to die, and their senescence can be initiated by environmental cues. As new leaves are initiated from the shoot apical meristem, older leaves often are shaded and lose the ability to function efficiently in photosynthesis. Senescence recovers a portion of the valuable resources that the plant invested in leaf formation. During senescence, hydrolytic enzymes break down many cellular proteins, carbohydrates, and nucleic acids. The component sugars, nucleosides, and amino acids are then transported back into the plant via the phloem, where they will be reused for synthetic processes. Many minerals also are transported out of senescing organs, back into the main body of the plant. Senescence of plant organs is frequently associated with abscission, a process whereby specific cells in the petiole differentiate to form an abscission layer, allowing the senescent organ to separate from the plant. The control of abscission by ethylene. The cytokinins and ethylene can act as signalling agents that regulate plant senescence.

Senescence occurs in a variety of organs and in response to many different cues. Many annual plants, including major crop plants such as wheat, maize, and soybeans, abruptly yellow and die following fruit production, even under optimal growing conditions. Senescence of the entire plant after a single reproductive cycle is called monocarpic senescence. Other types of senescence include the following:

- Senescence of aerial shoots in herbaceous perennials
- Seasonal leaf senescence (as in deciduous trees)
- Sequential leaf senescence (in which the leaves die when they reach a certain age)
- Senescence (ripening) of fleshy fruits; senescence of dry fruits
- Senescence of storage cotyledons and floral organs
- Senescence of specialized cell types (*e.g.*, trichomes, tracheids, and vessel elements)

The triggers for the various types of senescence are different and can be internal, as in monocarpic senescence, or external, such as day length and temperature in the autumnal leaf senescence of deciduous trees. Regardless of the initial stimulus, the different senescence patterns may share common internal programs in which a regulatory senescence gene initiates a cascade of secondary gene expression that eventually brings about senescence and death. Senescence is an ordered series of cytological and biochemical events Because it is genetically encoded, senescence follows a predictable course of cellular events. On the cytological level, some organelles are destroyed while others remain active. The chloroplast is the first organelle to deteriorate during the onset of leaf senescence, with the destruction of thylakoid protein components and stromal enzymes. In contrast to the rapid deterioration of chloroplasts, nuclei remain structurally and functionally intact until the late stages of senescence. Senescing tissues carry out cata-bolic processes that require the de novo synthesis of various hydrolytic enzymes, such as proteases, nucleases, lipases, and chlorophyll-degrading enzymes. The synthesis of these senescence-specific enzymes involves the activation of specific genes. Not surprisingly, the levels of most leaf mRNAs decline significantly during the senescence phase, but the abundance of certain specific mRNA transcripts increases. Genes whose expression decreases during senescence are called senescence down-regulated genes (SDGs). SDGs include genes that encode proteins involved in photosynthesis. However, senescence involves much more than the simple switching off of photosynthesis genes. Genes whose expression is induced during senescence are called senescence-associated genes (SAGs). SAGs include genes that encode hydrolytic enzymes, such as proteases, ribonucleases, and lipases, as well as enzymes involved in the biosynthesis of ethylene, such as ACC (laminocyclopropane-l-carboxylic acid) synthase and ACC oxidase. SAGs of another class have secondary functions in senescence. These genes encode enzymes involved in the conversion or remobilization of breakdown products, such as glutamine synthetase, which catalyzes the conversion of ammonium to glutamine and is responsible for nitrogen recycling from senescing tissues.

Programmed Cell Death is a Specialized Type of Senescence

Senescence can occur at the level of the whole plant, as in monocarpic senescence; at the organ level, as in leaf senescence; and at the cellular level, as in tracheary element differentiation. The process whereby individual cells activate an intrinsic senescence program is called programmed cell death (PCD). PCD plays an important part in animal development. PCD can be initiated by specific signals, such as errors in DNA replication during division, and involves the expression of a characteristic set of genes. The expression of these genes results in cell death. Much less is known about PCD in plants (Pennell and Lamb 1997). PCD in animals is usually accompanied by a distinct set of morphological and biochemical changes called apoptosis (plural apoptoses) (from a Greek word meaning "falling off," as in autumn leaves). During apoptosis, the cell nucleus condenses and the nuclear DNA fragments in a specific pattern caused by degradation of the DNA between nucleosomes. Some plant cells, particularly in senescing tissues, exhibit similar cytological changes. PCD also appears to occur during the differentiation of xylem tracheary elements,during which the nuclei and chromatin degrade and the cytoplasm disappears. These changes result fromthe activation of genes that encode nucleases and proteases. One of the important functions of PCD in plants is protection against pathogenic organisms. When a pathogenic organism infects a plant, signals from the pathogen cause the plant cells at the site of the infection to quickly accumulate high concentrations of toxic phenolic compounds and die. The dead cells form a small circular island of cell death called a necrotic lesion. The necrotic lesion isolates and prevents the infection from spreading to surrounding healthy tissues by surroundingthe pathogen with a toxic and nutritionally depleted environment.This rapid, localized cell death due to pathogen attack is called the hypersensitive response. The existence of Arabidopsis mutants that can mimic the effect of infection and trigger the entire cascade of events leading to the formation of necrotic lesions, even in the absence of the pathogen, has demonstrated that the hypersensitive response is a genetically programmed process rather than simple necrosis.

8.15. Plant Movements

The higher plants which are fixed to the substratum by means of roots need not move from one place of another,but their organs rae in a continuous state of movement.

Classification of Plant Movements

- **Plant Movements**
  - **Movement of Locomotion**
    - **Autonomic**
      - a. Ciliary
      - b. Amoeboid
      - c. Cyclosis
    - **Paratonic (Tactic)**
      - a. Phototactic
      - b. Chemotactic
      - c. Thermotactic
  - **Movement of Curvature**
    - **Growth Movement**
      - **Autonomic**
        - a. Nastic
        - b. Nutational
      - **Paratonic**
        - a. Phototropic
        - b. Geotropic
        - c. Hydrotropic
        - d. Thigmotropic
        - e. Chemotropic
    - **Variation Movement**
      - **Autonomic**
      - **Paratonic**
        - a. Nyctinastic
        - b. Seismonastic
        - c. Thigmonastic

Movement of Locomotion

1. Autonomic Movement of Locomotion

Those movement of locomotion which are not induced by external stimuli are called autonomic.

(i) Ciliary movements - Some motile forms of unicellular *e.g.* chlmydomonas, or multicellular *e.g.* volvox or the motile zoospore etc.

(ii) Amoeboid movement – Some lower forms of fungi consists of unicellular, nacked mass of protoplasm.

(iii) Cyclosis – It is the circulation of protoplasm in the cells. It is also known as protoplasmic streaming.

2. Paratonic Movements of Locomotion (Tactic movement)

The movement of locomotion which are brought about by some unidirectional external stimuli are known as tactic movements. Depending upon the nature of the stimuli these may be:

(i) Phototactic movement (Phototaxis) – The movement of the free living organism towards or away from light *e.g.* some free living algae. Phototaxis is also exhibited by chloroplast of the green plant cells.

(ii) Chemotactic movement (Chemotaxis) - The tactic movement in which external stimuli has the form of a gradient of chemical concentration, *e.g.* movement of antherozoid towards the neck of archegonia in bryophyte and pteridophyta.

(iii) Thermotactic movement (Thermotaxis) – The movement of free organism in response to the external stimuli of temperature, *e.g.*, Chlamydomonas cells move towards warmer water.

Movement of Curvature

1. Growth Movements

(A) Autonomic Growth Movement

(i) Nastic movements - These are the growth movements resulting from a difference between the rates of extension growth on opposite side of an organ.When the adaxial or morphologically upper side of an organ grows more rapidaly than the abaxial side,the resulting curvature is termed epinasty, *e.g.* in downward curling of leaf blade. When the abaxial side grow rapidaly than adaxial side,the movement is said to be hyponasty, *e.g.* in upward curling of leaf blade.

(ii) Nutational movement - Spiral course pursued by apex of a plant organ during growth due to continous change in position of most rapidaly growth of the region of the organ is termed as nutation. Such movements can not be seen by naked eyes because of slowness with which they occure.

Example of nutation is growth of apex in stem, roots, tendrils, flower stalks etc.

(B) Paratonic Growth Movements (Tropic or tropism)

Tropism is the movement of curvature brought about by more growth on one side and less gowth on opposite side of plant organ induced by some external stimuli. Depending upon the nature of stimuli the tropism can be phototropism,geotropism, hydrotropism, thigomotropism and chemotropism

Phototropism

The stem and other aerial parts of the plants bend towards the light source *i.e.*, they are positively phototropic, while roots and other underground organs bends away from the light source *i.e.*, they are are negatively phototropic. There are,however,some exceptions for example some tendrils and stems are negatively phototropic and some roots (coralloid roots of cycas) are positively phototropic. Leaves on the other hand are plageophototropic or transversely phototropic *i.e.*,they are placed at right angle to the direction of light. Similarly, some response change during the development of the plant. The stalk of overy in Arachis hypogeal is positively phototropic before fertilization but as soon as fertilization occurs it becomes negatively phototropic. This results the formation of fruits under ground.

The extreme apices are positive photoperception sites. However,in stem the maximum sensitive region lies about 0.1 – 0.2 inches below the apex. Blue light is most effective in producing phototropic curvature. However, U.V. light can also induce phototropism but to lesser extent

Mechanism of Phototropism

It has been well established that auxin plays an important role in causing phototropic curvature. The theory proposed by Cholodny –went is based on the role of auxin in phototropism.

The curvature brought about in the stem tip and root tip is due to the fact that one side of the tip grow more rapidaly as compared to opposite side. Auxin is responsible for unilateral growth of the tips.

Geotropism

The term is applied to growth movements induced by stimulus of gravity. Growth of an organ towards the centre of the earth is termed positive geotropism, *e.g.* main roots, and growth away from the centre of earth is termed as negative geotropism, *e.g.* main stem. When the axis of an organ grows at right angle to the direction of gravity it is said to be diageotropic, *e.g.* rhizome of certain grasses or stolons of potato, etc. Where an organ grows at an intermediate angle (*i.e.* between 0 to 90 or between 90 to 180 from the vertical) it is said to be plageotropic, *e.g.* leaves, lateral shoots, and lateral roots etc. Some of the lateral organ possess little geotropic sensitivity, they are called ageotropic.

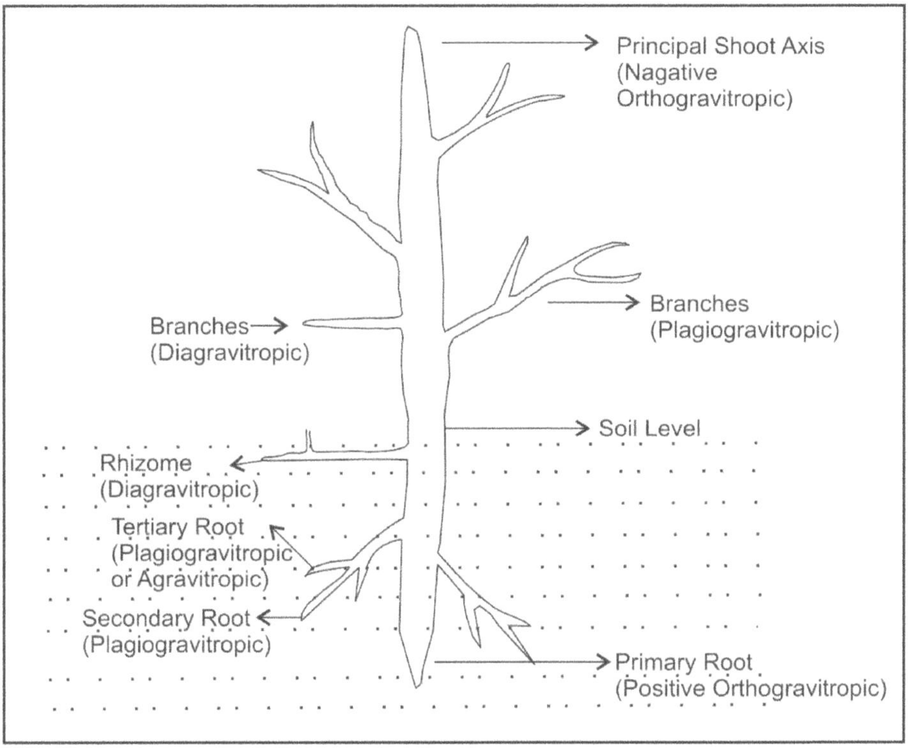

Figure 8.15: Movement of Curvature in Plants.

Geotropism stimulus is perceived mainly by the growing zones. In stem, the maximum sensitivity to graviperception occurs at the stem apex,where as, in case of roots the roots tip are sensitive to graviperception.

Mechanism of Geotropism

Many theory have been put forward to explain the mechanism of geotropism. Some are following:

1. Starch- Statolith Theory

The term statolith has been given to those cell inclusions which are sensitive to gravity (such as starch grains, crystal of calcium oxalate etc.). The cell which is gravity sensitive and also contains statolith has been termed statocyte or statocyst. Statolith freely move in statocyte.It has been suggested that graviperception is caused by internal sedimentation of statolith. The preesure exerted by sedimenting statolith upon cytoplasm against the lower lateral walls of the cell brings about the excitation. The excitation is then translated into metabolic changes. These changes caused unilateral growth resulting into geotropic curvature.

2. Cholodny–Went Theory

As in phototropism, the role of auxin is also in geotropism. When plant is placed in horizontal position its polar flow of auxin from tip to the the zone of elongation is disturbed. Higher auxin concentration is created on the lower side of stem and root tips than the upper side. Movement of auxin towards the direction of gravity has been demonstrated by various experiments.

Higher concentration of auxin promotes stem elongation on its lower half, whereas the growth in upper half becomes limited due to prescence of less auxin, thus, causing upward curvature (negative geotropism) in the stem. The case of root is different. Thus in case of roots the higher concentration of auxin in the lower part of the tip inhibits elongation growth. Less concentration of auxin, present in the upper half, promotes elongation growth with the result the root tip bends downward (positive geotropism). The cholodny-went model has been criticised because the shoots from which the apex has been removed can still show geocurvature.

Role of Calcium in Gravitropism

A role of calcium as aprimary effector of both coleoptiles and root gravi – curvature. Calcium is secreted by peripheral root cap cells into the surrounding mucilage. Secretion is uniform. Calium move lower side of the root cap and accumulate in and in elongation zone of the root. The accumulation is due to, 1. Calcium acts as a sink for auxin,thereby attracting auxin to lower side of root and 2. Calcium sensitizes the tissue to auxin.

Hydrotropism

The term hydrotropism is applied to growth movements in response to external stimulus of water. Roots are normally positively hydrotropic (*i.e.*, they bend towards the source of water). These curvature are again brought about by unilateral growth of the responding tips,although the exact regulation of hydrotropism is not well studied.

Thigmotropism (Haptotropism)

It is a movement in which a plant moves or grows in response to touch or contact stimuli. The prefix thigmo- comes from the Greek for "touch" (èéγµüò). Usually thigmotropism occurs when plants grow around a surface, such as a wall, pot, or trellis. Climbing plants, such as vines, develop tendrils that coil around supporting objects. Touched cells produce auxin and transport it to untouched cells. Some untouched cells will then elongate faster so cell growth bends around the object. Some seedlings also exhibit triple response, caused by pulses of ethylene which cause the stem to thicken (grow slower and stronger) and curve to start growing horizontally.

Roots also rely on touch to navigate their way through the soil. Generally, roots have a negative touch response, meaning when they feel an object, they would grow away from the object. This allows the roots to go through the soil with minimum resistance. Because of this behavior, roots are said to be negatively thigmotropic. Thigmotropism seems to be able to override the strong gravitropic

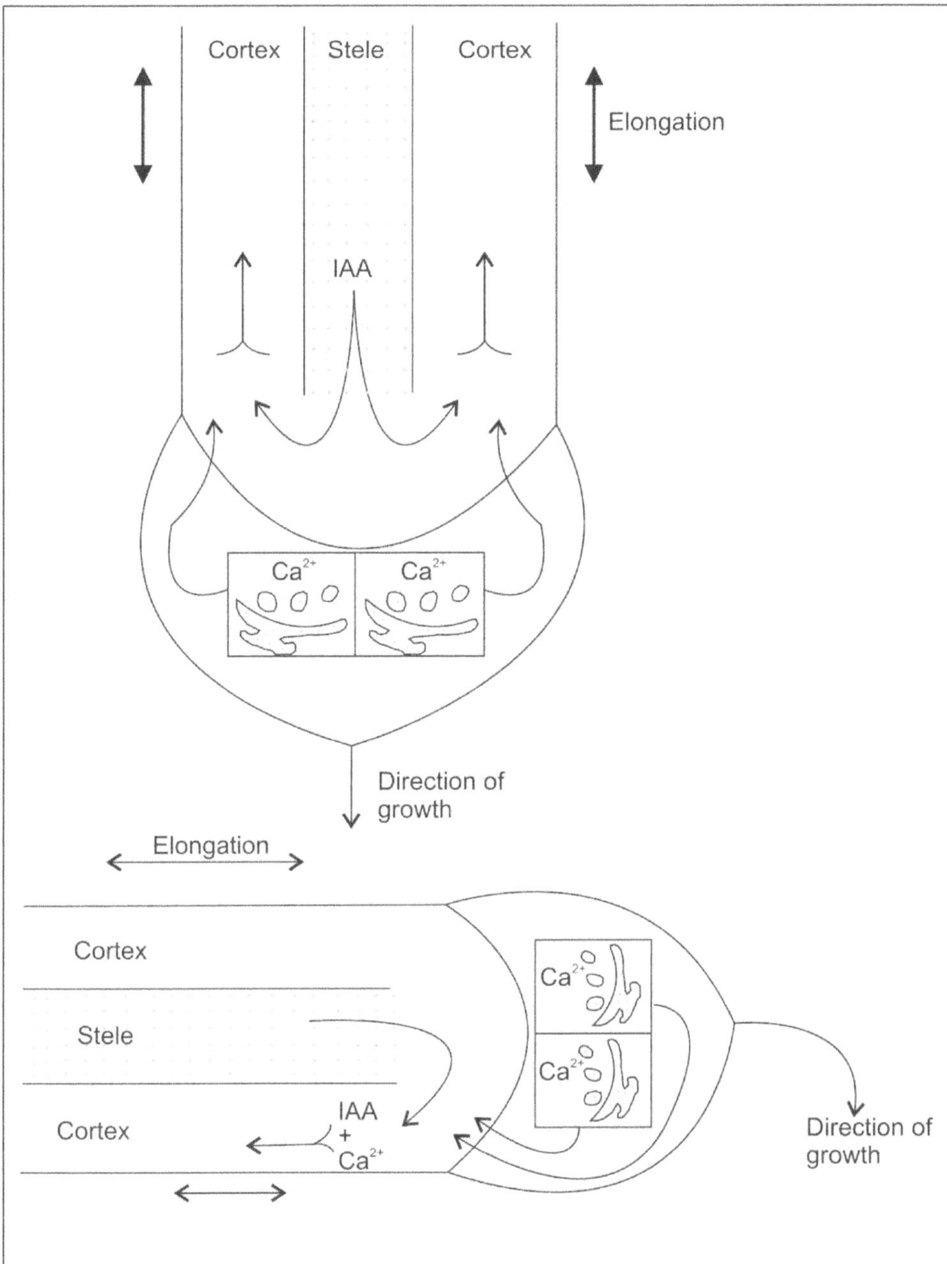

Figure 8.16: Role of Ca^{2+} in Gravitropism.

response of even primary roots. Charles Darwin performed experiments where he found that in a vertical bean root, a contact stimulus could divert the root away from the vertical. *Mimosa pudica* is well known for its rapid plant movement. The leaves close up and droop when touched. However, this is not a form of tropism, but a nastic movement, a similar phenomenon. The difference is that tropisms are influenced by the direction of their stimulus, while nastic movements are not.

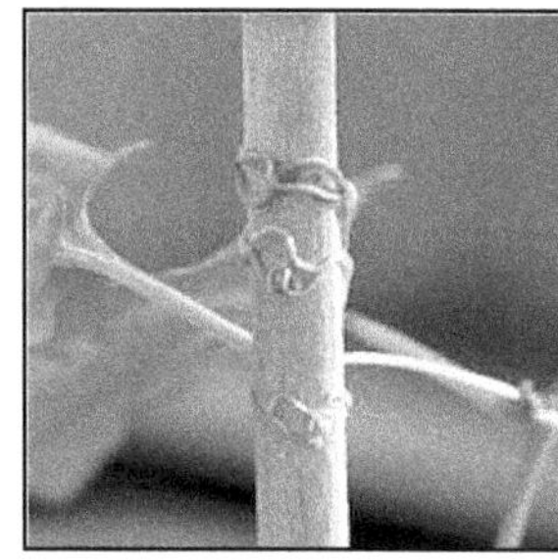

Figure 8.17: Thigmotropic Curvature.

Chemotropism

It is the growth of organisms (or parts of an organism, including individual cells) such as bacteria and plants, navigated by chemical stimulus from outside of the organism or organisms part. The response of the organism or organism part is termed 'positive' if the growth is towards the stimulus or 'negative' if the growth is away from the stimulus. An example of chemo-tropic movement can be seen during the growth of the pollen tube, where growth is always towards the ovules. It can be also written that conversion of flower into fruit is the example of chemo-tropism.Fertilization of flowers by pollen is achieved because the ovary releases chemicals that produce a positive chemotropic response from the developing pollen tube. An example of positive and negative chemotropism is shown by a plant's roots; the roots grow towards useful minerals displaying positive chemotropism, and grow away from harmful acids displaying negative chemotropism. Another example of chemotropic movement of includes the growth of individual neuronal cell axons in response to extra cellular signals, which guide the developing axon toinnervate the correct target tissue. Also, the addition of atmospheric nitrogen, also called nitrogen fixation, is an example of chemotropism.

Chemotropism is different from Chemotaxis, the major difference being that chemotropism is related to growth, while chemotaxis is related to locomotion.

3. Variation Movements

(A) Autonomic Variation Movements

The movement of variation, which is largly due to change in turgidity, when occurs without the external stimulus is known as autonomic variation movement. The best example of this is seen in Indian teligraph plant (*Desmodium gyrans*), in which the two lateral leaflets move up and down during the day time.

(B) Paratonic Variation Movements (Nastic movement)

Paratonic variation movement are determined by some external stimuli namely, light, temperature, chemicals and touch. Such movements mostly results from turgor changes, although some of them are accompanied with certain amount of growth. Nastic movement are stimulated by a change in the environmental conditions, and the direction of the response in determined by the organ, not by the stimulus. Nyctinastic and seismonastic responses depend on differential turgor movement in specialized motor organs,called the pulvinus.The pulvinus is bulbous structure most often found in plant families characterized by compound leaves, such as the leguminaceae and oxalidaceae. It occurs at the base of petiole (primary pulvinus), the pinna (secondary pulvinus), or the pinnule (tertiary pulvinus). The pulvinus contains a numbr of large,thin walled motor cells, which alter the position of the leaf by undergoing reversible changes in turgor.

(i) Nyctinastic Movements

The diurnal *i.e.*, change in day and night, movements of leaves and flower of some species which take up sleep position at night are called nyctinastic movement. Depending upon the stimulus they may be photonastic (light stimulated) or thermonastic (temperature stimulated).

The adaxial and abaxial sides of the pulvinus are also known as extensor and flexor region, respectively. Nyctinasty is the circadian rhythmic nastic movement of higher plants in response to the onset of darkness. Examples are the closing of the petals of a flower at dusk and the sleep movements of the leaves of manylegumes. Nyctinastic movements are controlled by the circadian clock and the light receptor phytochrome. Several leaf-opening and leaf-closing factors have been characterized biochemically. Anatomically, the movements are mediated by pulvini. In the SLEEPLESS mutation of *Lotus japonicus*, the pulvini are changed into petiole-like structures, rendering the plant incapable of closing its leaflets at night.

(ii) Chemonasty

When the external stimulus is some chemical substance, the variation movement is termed as chemonasty. *e.g.*, the long peripheral tentacles on sundew leaves respond to the prescence of organic nitrogenous compounds by bending towards the middle of the leaf. In another insectivorous plant – the butterwort (pinguicula), nitrogenous compounds causes the leaf to curl inwards on both the sides of the midrib.

(iii) Seismonastic (Haptonastic or thigmonastic) Movements

It is brought about in response to the external stimulus of touch. The best example of such movement is accomplished by so called ' sensitive plant" *Mimosa pudica*. Leaf of *Mimosa pudica* is compound with two or four pinnae, each of which bears numerous pairs of pinnules.The large primary and smaller secondary pulvini are present at the base of petiole and sub petioles, and still smaller tertiary pulvinus occurs at the base of each of the pinnules. When such a leaf is subjected to external stimulus, the main petiole droops, the pinnae come closer together and the pinnules turn upwards to come in contact in pairs.

Pulvinus is the swollen area consisting of large number of loosely packed parenchymatous cells separated by large intercellular spaces. The central portion of pulvinus is traversed by vascular strand. The cells on the upper half of the pulvinus have thick cell walls whereas those of lower portion contain thin walled cells.

If the terminal part of the leaf is touched the stimulus is transmitted to its base and then to other pinnules through the xylem sap, probably in the form of a hormone. The stimulusin the form of hormone can be transmitted to other leaves of the plants. Owing to the stimulation the turgor of cells of lower half of the pulvinus is lost, the cells of the upper half of pulvinus expand in longitudinal direction and whole pulvinus including central strand becomes curved and the petiole falls.

Several theories have been put forward to explain the mechanism by which turgor in the lower cells is lost. According to weintraub, the cytoplasmic contraction inthe cells results in ejaculation of fluid in to the intercellular space. Some of the osmotically active substance, probably potassium ions escape in to the intercellular space, which results into the concentration of the stretched cell walls to force out of the water of cell sap. After excitement is over, the osmotically active substance are believed to enter again in to the cell sap. This causes increases in the osmotic pressure of the cell sap in the lower half of the pulvinus. With the result cells absorb water from the intercellular spaces and become turgid. Simultaneously the leaf attains its normal position.

Some physiologist have been found that substances isolated from phloem sap of Mimosa and other species will stimulate closure of *Mimosa pulvini*, These active substances has been identified as a glycosylated derivatives of gallic acid, called " turgorin".

References

Abas, L. (2006). Intracellular trafficking and proteolysis of the Arabidopsis auxin-efflux facilitator PIN_2 are involved in root gravitropsis. *Nature Cell Biology* 8: 249-256.

Lin, C. (2002). Blue light receptors and signal transduction. *The Plant Cell.* Supplement 2002: S207-S225.

Trevaskis, B., Hemming, M.N., Dennis, E.S. and Peacock, W.J. (2007). The molecular basis of vernalization induced flowering in cereals. *Trends in Plant Sciences.* 12: 352-357.

Chapter 9

Stress Physiology

Plants are frequently exposed to environmental stresses. Stress is usually defined as an external factor that exerts a disadvantageous influence on the plant. This chapter will concern itself with environmental or abiotic factors that produce stress in plants, although biotic factors such as weeds, pathogens, and insect predation can also produce stress. In most cases, stress is measured in relation to plant survival, crop yield, growth (biomass accumulation), or the primary assimilation processes (CO_2 and mineral uptake), which are related to overall growth. The concept of stress is intimately associated with that of stress tolerance, which is the plant's fitness to cope with an unfavorable environment. In the literature the term *stress resistance* is often used interchangeably with *stress tolerance,* although the latter term is preferred. Note that an environment that is stressful for one plant may not be stressful for another. If tolerance increases as a result of exposure to prior stress, the plant is said to be acclimated (or hardened). Acclimation can be distinguished from adaptation, which usually refers to a genetically determined level of resistance acquired by a process of selection over many generations.

9.1. Water Deficity and Drought Resistance

Some drought resistance mechanisms, which are divided into several types. First we can distinguish between desiccation postponement (the ability to maintain tissue hydration) and desiccation tolerance (the ability to function while dehydrated), which are sometimes referred to as drought tolerance at high and low water potentials, respectively. The older literature often uses the term drought avoidance (instead of drought tolerance), but this term is a misnomer because drought is a meteorological condition that is tolerated by all plants that survive it and avoided by none. Athird category, drought escape, comprises plants that complete their life cycles during the wet season, before the onset of drought. These are the only true "drought avoiders." Among the desiccation postponers are water savers and water spenders. Water savers use water conservatively, preserving some in the soil for use late in their life cycle; water spenders aggressively consume water, often using prodigious quantities. The mesquite tree (*Prosopis* sp.) is an example of a water spender.

The water-limited productivity of plants depends on the total amount of water available and on thewater-use efficiency of the plant. A plant that is capable of acquiring more water or that has higher water-use efficiency will resist drought better. Some plants possess adaptations, such as the C4 and CAM modes of photosynthesis that allow them to exploit more arid environments. In addition, plants possess acclimation mechanisms that are activated in response to water stress. Water deficit can be defined as any water content of a tissue or cell that is below the highest water content exhibited at the most hydrated state. When water deficit develops slowly enough to allow changes in developmental processes, water stress has several effects on growth, one of which is a limitation in leaf expansion. Leaf area is important because photosynthesis is usually proportional to it. However, rapid leaf expansion can adversely affect water availability. If precipitation occurs only during winter and spring, and summers are dry, accelerated early growth can lead to large leaf areas, rapid water depletion, and too little residual soil moisture for the plant to complete its life cycle. In this situation, only plants that have some water available for reproduction late in the season or that complete the life cycle quickly, before the onset of drought (exhibiting drought escape), will produce seeds for the next generation. Either strategy will allow some reproductive success. The situation is different if summer rainfall is significant but erratic. In this case, a plant with large leaf area, or one capable of developing large leaf area very quickly, is better suited to take advantage of occasional wet summers. One acclimation strategy in these conditions is a capacity for both vegetative growth and flowering over an extended period. Such plants are said to be *indeterminate* in their growth habit, in contrast to *determinate* plants, which develop preset numbers of leaves and flower over only very short periods. Typically, as the water content of the plant decreases, its cells shrink and the cell walls relax. This decrease in cell volume results in lower turgor pressure and the subsequent concentration of solutes in the cells. The plasma membrane becomes thicker and more compressed because it covers a smaller area than before. Because turgor reduction is the earliest significant biophysical effect of water stress, turgor-dependent activities such as leaf expansion and root elongation are the most sensitive to water deficits. The decrease in turgor causes a decrease in growth rate. Inhibition of cell expansion results in a slowing of leaf expansion early in the development of water deficits. The smaller leaf area transpires less water, effectively conserving a limited water supply in the soil over a longer period. Reduction in leaf area can thus be considered a first line of defense against drought. In indeterminate plants, water stress limits not only leaf size, but also leaf number, because it decreases both the number and the growth rate of branches.

The total leaf area of a plant (number of leaves × surface area of each leaf) does not remain constant after all the leaves have matured. If plants become water stressed after a substantial leaf area has developed, leaves will senesce and eventually fall off. Abscission during water stress results largely from enhanced synthesis of and responsiveness to the endogenous plant hormone ethylene.

Mild water deficits also affect the development of the root system. Root-to-shoot biomass ratio appears to be governed by a functional balance between water uptake by the root and photosynthesis by the shoot. Simply stated, a shoot will grow until it is so large that water uptake by the roots becomes limiting to further growth; conversely, roots will grow until their demand for photosynthate from the shoot equals the supply. This functional balance is shifted if the water supply decreases.

Under these stress conditions, stomata closure reduces evaporation from the existing leaf area. Thus, stomatal closure can be considered a third line of defense against drought. Uptake and loss of water in guard cells changes their turgor and modulates stomatal opening and closing. Because guard cells are located in the leaf epidermis, they can lose turgor as a result of a direct loss of water by evaporation to the atmosphere. The decrease in turgor causes stomatal closure by hydropassive closure. This closing mechanism is likely to operate in air of low humidity, when direct water loss from the guard cells is too rapid to be balanced by water movement into the guard cells from adjacent epidermal cells. A second mechanism, called hydroactive closure, closes the stomata when the whole leaf or the roots are dehydrated and depends on metabolic processes in the guard cells. A reduction in the solute content of the guard cells results in water loss and decreased turgor, causing the stomata to close; thus the hydraulic mechanism of hydroactive closure is a reversal of the mechanism of stomatal opening. However, the control of hydroactive closure differs in subtle but important ways from stomatal opening. Solute loss from guard cells can be triggered by a decrease in the water content of the leaf, and abscisic acid (ABA) plays an important role in this process. Abscisic acid is synthesized continuously at a low rate in mesophyll cells and tends to accumulate in the chloroplasts. Besides ABA (Sauter *et al.*, 2001), other signals, such as pH and inorganic ion redistribution, appear to play a role in long-distance signaling between the roots and the shoots.

Water stress decreases both photosynthesis and the consumption of assimilates in the expanding leaves. As a consequence, water stress indirectly decreases the amount of photosynthate exported from leaves. Because phloem transport depends on turgor, decreased water potential in the phloem during stress may inhibit the movement of assimilates.

Osmotic adjustment, or accumulation of solutes by cells, is a process by which water potential can be decreased without an accompanying decrease in turgor or decrease in cell volume. Osmotic adjustment is a net increase in solute content per cell that is independent of the volume changes that result from loss of water. The decrease in Ys is typically limited to about 0.2 to 0.8 MPa, except in plants adapted to extremely dry conditions. Most of the adjustment can usually be accounted for by increases in concentration of a variety of common solutes, including sugars, organic acids, amino acids, and inorganic ions (especially K^+).

When a soil dries, its resistance to the flow of water increases very sharply, particularly near the *permanent wilting point*. At the permanent wilting point (usually about –1.5 MPa), plants cannot regain turgor pressure even if all transpiration stops.

Response to water stress is the production of a thicker cuticle that reduces water loss from the epidermis (cuticular transpiration). Although waxes are deposited in response to water deficit both on the surface

and within the cuticle inner layer, the inner layer may be more important in controlling the rate of water loss in ways that are more complex than by just increasing the amount of wax present.

The accumulation of compatible solutes in response to osmotic stress requires the activation of the metabolic pathways that biosynthesize these solutes. Several genes coding for enzymes associated with osmotic adjustment are turned on (up-regulated) by osmotic stress and/or salinity, and cold stress. These genes encode enzymes such as the following (Buchanan *et al.*, 2000): Δ 2 ?-Pyrroline-5-carboxylate synthase, a key enzyme in the proline biosynthetic pathway, Betaine aldehyde dehydrogenase, an enzyme involved in glycine betaine accumulation, *myo*-Inositol 6-*O*-methyltransferase, a rate-limiting enzyme in the accumulation of the cyclic sugar alcohol called pinitol Several other genes that encode well-known enzymes are

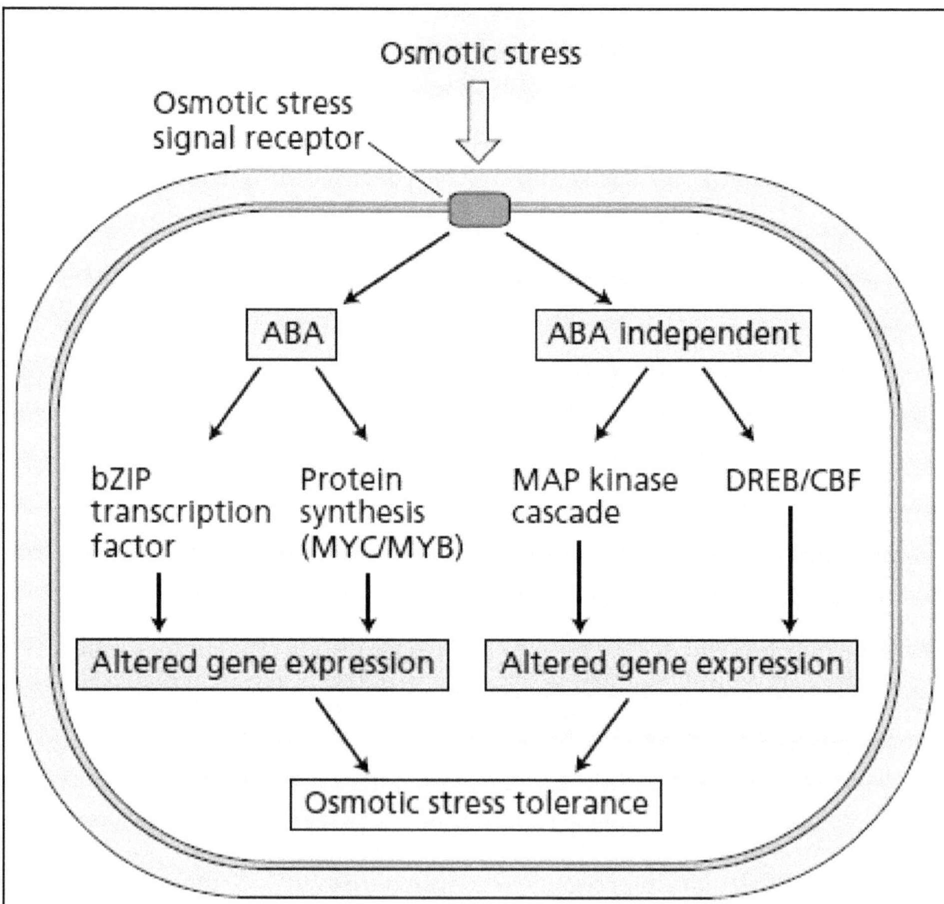

Figure 9.1: Signal Transduction for Osmotic Stress in Plant Cells.

induced b.A large group of genes that are regulated by osmotic stress was discovered by examination of naturally desiccating embryos during seed maturation. These genes code for so-called LEA proteins (named for *l*ate *e*mbryogenesis *a*bundant), and they are suspected to play a role in cellular membrane protection.y osmotic stress. Osmotic stress typically leads to the accumulation of ABA, so it is not surprising that products of ABA-responsive genes accumulate during osmotic stresses. However, not all genes that are up-regulated by osmotic stresses are ABAregulated. the promoters of several stress-induced gene have led to the identification of specific regulatory sequences for genes involved in different stresses. For example, the *RD29* gene contains DNA sequences that can be activated by osmotic stress, by cold, and by ABA(Yamaguchi-Shinozaki and Shinozaki 1994; Stockinger *et al.*, 1997). The promoters of ABA-regulated genes contain a sixnucleotide sequence element referred to as the ABA response element (ABRE), which probably binds transcriptional factors involved in ABA-regulated gene activation. The promoters of these genes, which are regulated by osmotic stress in an ABA-dependent manner,contain an alternative nine-nucleotide regulatory sequence element, the dehydration response element (DRE) which is recognized by an alternative set of proteins regulating transcription. Thus the genes that are regulated by osmotic stresses appear to be regulated either by signal transduction pathways mediated by the action of ABA (ABA-dependent genes), or by an ABA-independent, osmotic stress–responsive signal transduction pathway.

9.2. Heat Stress and Heat Shock

Most tissues of higher plants are unable to survive extended exposure to temperatures above 45°C. Nongrowing cells or dehydrated tissues (*e.g.*, seeds and pollen) can survive much higher temperatures than hydrated, vegetative, growing cells. Actively growing tissues rarely survive temperatures above 45°C, but dry seeds can endure 120°C, and pollen grains of some species can endure 70°C. In general, only single-celled eukaryotes can complete their life cycle at temperatures above 50°C, and only prokaryotes can divide and grow above 60°C. Periodic brief exposure to sublethal heat stresses often induces tolerance to otherwise lethal temperatures, a phenomenon referred to as induced thermotolerance. The water and temperature stress are interrelated; shoots of most C3 and C4 plants with access to abundant water supply are maintained below 45°C by evaporative cooling; if water becomes limiting, evaporative cooling decreases and tissue temperatures increase. Emerging seedlings in moist soil may constitute an exception to this general rule. These seedlings may be exposed to greater heat stress than those in drier soils because wet, bare soil is typically darker and absorbs more solar radiation than drier soil.

High Leaf Temperature and Water Deficit Lead to Heat Stress

Many CAM, succulent higher plants, such as *Opuntia* and *Sempervivum*, are adapted to high temperatures and can tolerate tissue temperatures of 60 to 65°C under conditions of intense solar radiation in summer. Because CAM plants keep their stomata closed during the day, they cannot cool by transpiration. Instead, they dissipate the heat from incident solar radiation by re-emission of longwave (infrared) radiation and loss of heat by conduction and convection. On the other hand, typical, nonirrigated C3 and C4 plants rely on transpirational cooling to lower leaf temperature. In these plants, leaf temperature can readily rise 4 to 5°C above ambient air temperature in bright sunlight near midday, when soil water deficit causes partial stomatal closure or when high relative humidity reduces the potential for evaporative cooling. Increases in leaf temperature during the day can be pronounced in plants from arid and semiarid regions experiencing drought and high irradiance from sunshine. Heat stress is also a potential danger in greenhouses, where low air speed and high humidity decrease the rate of leaf cooling. A moderate degree of heat stress slows growth of the whole plant. Some irrigated crops, such as cotton, use transpirational cooling to dissipate heat. In irrigated cotton, enhanced transpirational cooling is associated with higher agronomic yields.

At High Temperatures, Photosynthesis is Inhibited before Respiration

Both photosynthesis and respiration are inhibited at high temperatures, but as temperature increases, photosynthetic rates drop before respiratory rates. The temperature at which the amount of CO_2 fixed by photosynthesis, equals the amount of CO_2 released by respiration, in a given time interval is called the temperature compensation point. At temperatures above the temperature compensation point, photosynthesis cannot replace the carbon used as a substrate for respiration. As a result, carbohydrate reserves decline, and fruits and vegetables lose sweetness. This imbalance between photosynthesis and respiration is one of the main reasons for the deleterious effects of high temperatures. In the same plant the temperature compensation point is usually lower for shade leaves than for sun leaves that are exposed to light (and heat). Enhanced respiration rates relative to photosynthesis at high temperatures are more detrimental in C3 plants than in C4 or CAM plants because the rates of both dark respiration and photorespiration are increased in C3 plants at higher temperatures.

The high temperature limits for photosynthesis is generally marked by an abrupt increase in chlorophyll fluorescence. This fluorescence rise reflects irreversible daage to the energy transfer mechanism of PSII. The result is that an absorbed energy cannot be used photochemically by PS II and must be reradiated as light.

High Temperature Reduces Membrane Stability

The stability of various cellular membranes is important during high-temperature stress, just as it is during chilling and freezing. Excessive fluidity of membrane lipids at high temperatures is correlated with loss of physiological function. In oleander (*Nerium oleander*), acclimation to high temperatures is associated with a greater degree of saturation of fatty acids in membrane lipids, which makes the membranes less fluid (Raison *et al.*, 1982). At high temperatures there is a decrease in the strength of hydrogen bonds and electrostatic interactions between polar groups of proteins within the aqueous phase of the membrane. High temperatures thus modify membrane composition and structure and can cause leakage of ions. Membrane disruption also causes the inhibition of processes such as photosynthesis and respiration that depend on the activity of membrane associated electron carriers and enzymes.

Both drought resistance and heat resistance depend on the same adaptations: reflective leaf hairs and leaf waxes; leaf rolling and vertical leaf orientation; and growth of small, highly dissected leaves to minimize the boundary layer thickness and thus maximize convective and conductive heat loss. Some desert shrubs–for example, white brittlebush (*Encelia farinosa*, family Compositae)–have dimorphic leaves to avoid excessive heating: Green, nearly hairless leaves found in the winter are replaced by white, pubescent leaves in the summer.

At Higher Temperatures, Plants Produce Heat Shock Proteins

In response to sudden, 5 to 10°C rises in temperature, plants produce a unique set of proteins referred to as heat shock proteins (HSPs). Most HSPs function to help cells withstand heat stress by acting as molecular chaperones. Heat stress causes many cell proteins that function as enzymes or structural components to become unfolded or misfolded, thereby leading to loss of proper enzyme structure and activity. Such misfolded proteins often aggregate and precipitate, creating serious problems within the cell. HSPs act as molecular chaperones and serve to attain a proper folding of misfolded, aggregated proteins and to prevent misfolding of proteins. This facilitates proper cell functioning at elevated, stressful temperatures. Heat shock proteins were discovered in the fruit fly (*Drosophila melanogaster*) and have since been identified in other animals, and in humans, as well as in plants, fungi, and microorganisms. For example, when soybean seedlings are suddenly shifted from 25 to 40°C (just below the lethal temperature), synthesis of the set of mRNAs and proteins commonly found in the cell is suppressed, while transcription and translation of a set of 30 to 50 other proteins (HSPs) is enhanced. New mRNA transcripts for HSPs can be detected 3 to 5 minutes after heat shock. Different HSPs are localized to the nucleus, mitochondria, chloroplasts, endoplasmic reticulum, and cytosol. Members of the HSP60, HSP70, HSP90, and HSP100 groups act as molecular chaperones, involving ATP-dependent stabilization and folding of proteins, and the assembly of oligomeric proteins. Some HSPs assist in polypeptide transport across membranes into cellular compartments.

In addition, heat stress can alter the rate of metabolic reactions that consume or produce protons, and it can affect the activity of proton-pumping ATPases that pump protons from the cytosol into the apoplast or vacuoles. This might lead to an acidification of the cytosol, which could cause additional metabolic perturbations during stress. Cells can have metabolic acclimation mechanisms that ameliorate these effects of heat stress on metabolism. One of the metabolic acclimations to heat stress is the accumulation of the nonprotein amino acid γ-aminobutyric acid (GABA). During episodes of heat stress, GABA accumulates to levels six- to tenfold higher than in unstressed plants. GABA is synthesized from the amino acid L-glutamate, in a single reaction catalyzed by the enzyme glutamate decarboxylase (GAD). GAD is one of several enzymes whose activity is modulated by the calcium-activated, regulatory protein calmodulin

9.3. Chilling and Freezing

Chilling temperatures are too low for normal growth but not low enough for ice to form. Typically, tropical and subtropicalspecies are susceptible to chilling injury. Among crops, maize, *Phaseolus* bean, rice, tomato, cucumber, sweet potato, and cotton are chilling sensitive. *Passiflora, Coleus,* and *Gloxinia* are examples of susceptible ornamentals. When plants growing at relatively warm temperatures (25 to 35°C) are cooled to 10 to 15°C, chilling injury occurs: Growth is slowed, discoloration or lesions appear on leaves, and the foliage looks soggy, as if soaked in water for a long time. If roots are chilled, the plants may wilt. Species that are generally sensitive to chilling can show appreciable variation in their response to chilling temperatures. Genetic adaptation to the colder temperatures associated with high altitude improves chilling resistance. In addition, resistance often increases if plants are first hardened (acclimated) by exposure to cool, but noninjurious, temperatures. Chilling damage thus can be minimized if exposure is slow and gradual. Sudden exposure to temperatures near 0°C, called *cold shock,* greatly increases the chances of injury. Freezing injury, on the other hand, occurs at temperatures below the freezing point of water. Full induction of tolerance to freezing, as with chilling, requires a period of acclimation at cold temperatures.

Membrane Properties Change in Response to Chilling Injury

Leaves from plants injured by chilling show inhibition of photosynthesis, slower carbohydrate translocation, lower respiration rates, inhibition of protein synthesis, and increased degradation of existing proteins. All of these responses appear to depend on a common primary mechanism involving loss of membrane function during chilling. For instance, solutes leak from the leaves of chilling sensitive *Passiflora maliformis* (conch apple) floated on water at 0°C, but not from those of chilling-resistant *Passiflora caerulea* (passion flower). Loss of solutes to the water reflects damage to the plasma membrane and possibly also to the tonoplast. In turn, inhibition of photosynthesis and of respiration reflects injury to chloroplast and mitochondrial membranes. Why are membranes affected by chilling? Plant membranes consist of a lipid bilayer interspersed with proteins and sterols. The physical properties of the lipids greatly influence the activities of the integral membrane proteins, including H^+-ATPases, carriers, and channel-forming proteins that regulate the transport of ions and other solutes, as well as the transport of enzymes on which metabolism depends. In chilling-sensitive plants, the lipids in the bilayer have a high percentage of saturated fatty acid chains, and membranes with this composition tend to solidify into a semicrystalline state at a temperature well above 0°C. Keep in mind that saturated fatty acids that have no double bonds and lipids containing *trans*-monounsaturated fatty acids solidify at higher temperatures than do membranes composed of lipids that contain unsaturated fatty acids.

As the membranes become less fluid, their protein components can no longer function normally. The result is inhibition of H^+-ATPase activity, of solute transport into and out of cells, of energy transduction, and of enzyme-dependent metabolism. In addition, chilling- sensitive leaves exposed to high photon fluxes and chilling temperatures are photoinhibited, causing acute damage to the photosynthetic machinery. Membrane lipids from chilling-resistant plants often have a greater proportion of unsaturated fatty acids than those from chilling-sensitive plants, and during acclimation to cool temperatures the activity of desaturase enzymes increases and the proportion of unsaturated lipids rises (Williams *et al.*, 1988; Palta *et al.*, 1993). This modification lower the temperature at which membrane lipids begin a gradual phase change from fluid to semicrystalline and allows membranes to remain fluid at lower temperatures. Thus, desaturation of fatty acids provides some protection against damage from chilling.

Ice Crystal Formation and Protoplast Dehydration Kill Cells

The ability to tolerate freezing temperatures under natural conditions varies greatly among tissues. Fully hydrated, vegetative cells can also retain viability if they are cooled very quickly to avoid the formation of large, slow-growing ice crystals that would puncture and destroy subcellular structures. Ice crystals that form during very rapid freezing are too small to cause mechanical damage. Conversely, rapid warming of frozen tissue is required to prevent the growth of small ice crystals into crystals of a damaging size, or to prevent loss of water vapor by sublimation. Ice usually forms first within the intercellular spaces, and in the xylem vessels, along which the ice can quickly propagate. This ice formation is not lethal to hardy plants, and the tissue recovers fully if warmed. However, when plants are exposed to freezing temperatures for an extended period, the growth of extracellular ice crystals results in the movement of liquid water from the protoplast to the extracellular ice, causing excessive dehydration. Several hundred molecules are needed for an ice crystal to begin forming. The process whereby these hundreds of water molecules start to form a stable ice crystal is called ice nucleation.

Sugars and some of the cold-induced proteins are suspected to have cryoprotective (cryo- = "cold") effects; they stabilize proteins and membranes during dehydration induced by low temperature. In winter wheat, the greater the sucrose concentration, the greater the freezing tolerance. During cold acclimation of winter cereals, soluble sugars accumulate in the cell walls, where they may help restrict the growth of ice. A cryoprotective glycoprotein has been isolated from leaves of cold-acclimated cabbage (*Brassica oleracea*). In vitro, the protein protects thylakoids isolated from nonacclimated spinach (*Spinacia oleracea*) against damage from freezing and thawing.

In seedlings of alfalfa (*Medicago sativa* L.), tolerance to freezing at –10°C is greatly improved by previous exposure to cold (4°C) or by treatment with exogenous ABA without exposure to cold. These treatments cause changes in the pattern of newly synthesized proteins. Protein synthesis is necessary for the development of freezing tolerance, and several distinct proteins accumulate during acclimation to cold, as a result of changes in gene expression (Guy, 1999). Isolation of the genes for these proteins reveals that several of the proteins that are induced by low temperature share homology with the RAB/LEA/DHN (responsive to ABA, late embryo abundant, and dehydrin, respectively) protein family.

9.4. Salinity Stress

Under natural conditions, terrestrial higher plants encounter high concentrations of salts close to the seashore and in estuaries where seawater and freshwater mix or replace each other with the tides. Far inland, natural salt seepage from geologic marine deposits can wash into adjoining areas, rendering them unusable for agriculture. However, a much more extensive problem in agriculture is the accumulation of salts from irrigation water. Evaporation and transpiration remove pure water (as vapor) from the soil, and this water loss concentratessolutes in the soil. When irrigation water contains a high concentration of solutes and when there is no opportunity to flush out accumulated salts to a drainage system, salts can quickly reach levels that are injurious to salt-sensitive species. It is estimated that about one-third of the irrigated land on Earth is affected by salt.

Salt Accumulation in Soils Impairs Plant Function and Soil Structure

In discussing the effects of salts in the soil, we distinguish between high concentrations of Na^+, referred to as sodicity, and high concentrations of total salts, referred to as salinity. The two concepts are often related, but in some areas Ca^{2+}, Mg^{2+}, and SO^{2-}_4, as well as NaCl, can contribute substantially to salinity. The high Na^+ concentration of a sodic soil can not only injure plants directly but also degrade the soil structure, decreasing porosity and water permeability. A sodic clay soil known as caliche is so hard and impermeable that dynamite is sometimes required to dig through it! In the field, the salinity of soil water or irrigation water is measured in terms of its electrical conductivity or in terms of osmotic potential. Pure water is a very poor conductor of electric current; the conductivity of a water sample is due to the ions dissolved in it. The higher the salt concentration in water, the greater its electrical conductivity and the lower its osmotic potential (higher osmotic pressure). The quality of irrigation water in semiarid and arid regions is often poor.

Salinity Depresses Growth and Photosynthesis in Sensitive Species

Plants can be divided into two broad groups on the basis of their response to high concentrations of salts. Halophytes are native to saline soils and complete their life cycles in that environment. Glycophytes (literally "sweet plants"), or nonhalophytes, are not able to resist salts to the same degree as halophytes. Usually there is a threshold concentration of salt above which glycophytes begin to show signs of growth inhibition, leaf discoloration, and loss of dry weight. Among crops, maize, onion, citrus, pecan, lettuce, and bean are highly sensitive to salt; cotton and barley are moderately tolerant; and sugar beet and date palms are highly tolerant (Greenway and Munns 1980). Some species that are highly tolerant of salt, such as *Suaeda maritima* (a salt marsh plant) and *Atriplex nummularia.*

Salt Injury Involves Both Osmotic Effects and Specific Ion Effects

Dissolved solutes in the rooting zone generate a low (more negative) osmotic potential that lowers the soil water potential. The general water balance of plants is thus affected because leaves need to develop an even lower water potential to maintain a "downhill" gradient of water potential between the soil and the leaves. This effect of dissolved solutes is similar to that of a soil water deficit, and most plants respond to excessive levels of soil salinity in the same way.In addition to the plant responses to low water potential, specific ion toxicity effects also occur when injurious concentrations of ions–particularly Na^+, Cl^-, or SO_4^{2-} accumulate in cells. Photosynthesis is inhibited when high concentrations of Na^+ and/or Cl^- accumulate in chloroplasts. Since photosynthetic electron transport appears relatively insensitive to salts, either carbon metabolism or photophosphorylation may be affected.

Plants Use Different Strategies to Avoid Salt Injury

In plants that are salt sensitive, resistance to moderate levels of salinity in the soil depends in part on the ability of the roots to prevent potentially harmful ions from reaching the shoots. The Casparian strip imposes a restriction to the movements of ions into the xylem. To bypass the Casparian strips, ions need to move from the apoplast to the symplastic pathway across cell membranes. This transition offers salt-resistant plants a mechanism to partially exclude harmful ions. Sodium ions enter roots passively (by moving down an electrochemical-potential gradient, so root cells must use energy to extrude Na^+ actively back to the outside solution. By contrast, Cl^- is excluded by negative electric potential across the cell membrane, and the low permeability of root plasma membranes to this ion. Movement of Na^+ into leaves is further minimized by absorption of Na^+ from the transpiration stream (xylem sap) during its movement from roots to shoots and leaves. Some salt-resistant plants, such as salt cedar (*Tamarix* sp.) and salt bush (*Atriplex* sp.), do not exclude ions at the root, but instead have salt glands at the surface of the leaves. The ions are transported to these glands, where the salt crystallizes and is no longer harmful. In general, halophytes have a greater capacity than glycophytes for ion accumulation in shoot cells.

Ion Exclusion Is Critical for Acclimation and Adaptation to Salinity Stress

In terms of metabolic energy, use of ions to balance tissue water potential in a saline environment clearly has a lower energy cost for the plant than use of carbohydrates or amino acids, the production of which has a significantly higher energy cost. On the other hand, high ion concentrations are toxic to many cytosolic enzymes, so ions must be accumulated in the vacuole to minimize toxic concentrations in the cytosol. Because NaCl is the most abundant salt encountered by plants under salinity stress, transport systems that facilitate compartmentation of Na^+ into the vacuole are critical (Binzel *et al.*, 1988). Both Ca^{2+} and K^+ affect intracellular Na^+ concentrations (Zhong and Läuchli 1994). At high concentrations of Na^+, K^+ uptake through a high-affinity K^+–Na^+ transporter, HKT1, is inhibited, and the transporter operates as an Na^+ uptake system. Calcium, on the other hand, enhances K^+/Na^+ selectivity and in so doing increases salt tolerance.

9.5. Oxygen Deficiency

Roots usually obtain sufficient oxygen (O_2) for aerobic respiration directly from the gaseous space in the soil. Gas-filled pores in well-drained, well-structured soil readily permit the diffusion of gaseous O_2 to depths of several meters. Consequently, the O_2 concentration deep in the soil is similar to that in humid air. However, soil can become flooded or waterlogged when it is poorly drained or when rain or irrigation is excessive. Water then fills the pores and blocks the diffusion of O_2 in the gaseous phase. Dissolved oxygen diffuses so slowly in stagnant water that only a few centimeters of soil near the surface remain oxygenated. When temperatures are low and plants are dormant, oxygen depletion is very slow and the consequences are relatively harmless. However, when temperatures are higher (greater than 20°C), oxygen consumption by plant roots, and soil fauna and microorganisms, can totally deplete the oxygen from the bulk of the soil water in as little as 24 hours. Flooding-sensitive plants are severely damaged by 24 hours of anoxia. The growth and survival of many plant species are greatly depressed under such conditions, and crop yields can be severely reduced. For example, garden pea (*Pisum sativum*) yields can be halved by 24 hours of flooding, making garden pea an example of a flooding-sensitive plant. Other plants, particularly species not adapted to grow in continually wet conditions and many crop plants, are affected by flooding in a milder way and are considered flooding-tolerant plants. Flooding-tolerant plants can withstand anoxia (lack of oxygen) temporarily, but not for prolonged periods of more than a few days. On the other hand, specialized natural vegetation found in wetlands such as marshes and swamps, and crops such as rice, are well adapted to resist oxygen deficiency in the root environment. Wetland plants can resist anoxia, and they grow and survive for extended periods of up to months with their root systems in anoxic conditions. Practically all plants require oxygen when they are engaging in rapid metabolic activity, and plants can be classified according to the time they can withstand anoxic conditions in their root environment without demonstrating substantial damage.

Anaerobic Microorganisms are Active in Water-Saturated Soils

When soil is completely depleted of molecular O_2, the function of soil microbes becomes significant for plant life and growth. Anaerobic soil microorganisms (anaerobes) derive their energy from the reduction of nitrate (NO_3^-) to nitrite (NO_2^-) or to nitrous oxide (N_2O) and molecular nitrogen (N_2). These gases (N_2O and N_2) are lost to the atmosphere in a process called denitrification. As conditions become more reducing, anaerobes reduce Fe_3^+ to Fe_3^+, and because of its greater solubility, Fe_2^+ can rise to toxic concentrations when some soils are anaerobic for many weeks. Other anaerobes may reduce sulfate (SO_4^{2-}) to hydrogen sulphide (H_2S), which is a respiratory poison. When anaerobes have an abundant supply of organic substrate, bacterial metabolites such as acetic acid and butyric acid are released into the soil water, and these acids along with reduced sulfur compounds account for the unpleasant odor of waterlogged soil. All of these substances made by microorganisms under anaerobic conditions are toxic to plants at high concentrations.

Roots are Damaged in Anoxic Environments

Root respiration rate and metabolism are affected even before O_2 is completely depleted from the root environment. The critical oxygen pressure (COP) is the oxygen pressure at which the respiration rate is first slowed by O_2 deficiency. In older zones of the root, where cells are mature and fully vacuolated and the respiration rate is lower, the COP is often in the range of 0.1 to 0.05 atmosphere. When O_2 concentrations are below the COP, the center of the root becomes *anoxic* (completely lacking oxygen) or *hypoxic* (partly deficient in oxygen). The COP is lower when respiration slows down at cooler temperatures; it also depends on how bulky the organ is and how tightly the cells are packed. Large, bulky fruits are able to remain fully aerobic because of the large intercellular spaces that readily allow gaseous diffusion. For single cells, an O_2 partial pressure as low as 0.01 atmosphere (1 per cent O_2 in the gaseous phase) can be adequate because diffusion over short distances ensures an adequate O_2 supply to mitochondria. Avery low partial pressure of O_2 at the mitochondrion is sufficient to maintain oxidative phosphorylation. The large difference between the COP values for an organ or tissue and the O_2 requirements of mitochondria is explained by the slow diffusion of dissolved O_2 in aqueous media. In the absence of O_2, electron transport and oxidative phosphorylation in mitochondria cease, the tricarboxylic acid cycle cannot operate, and ATP can be produced only by fermentation. Thus when the supply of O_2 is insufficient for aerobic respiration, roots first begin to ferment pyruvate to lactate, through the action of lactate dehydrogenase (LDH).

Damaged O_2-Deficient Roots Injure Shoots

Anoxic or hypoxic roots lack sufficient energy to support physiological processes on which the shoots depend. Experiments have shown that the failure of the roots of wheat or barley to absorb nutrient ions and transport them to the xylem (and from there to the shoot) quickly leads to a shortage of ions within developing and expanding tissues. Older leaves senesce prematurely because of reallocation of phloem-mobile elements (N, P, K) to younger leaves. The lower permeability of roots to water often leads to a decrease in leaf water potential and wilting, although this decrease is temporary if stomata close, preventing further water loss by transpiration. Hypoxia also accelerates production of the ethylene precursor ACC (1-aminocyclopropane-1-carboxylic acid) in roots. In tomato, ACC travels via the xylem sap to the shoot, where, in contact with oxygen, it is converted by ACC oxidase to ethylene. The upper (adaxial) surfaces of the leaf petioles of tomato and sunflower have ethylene-responsive cells that expand more rapidly when ethylene concentrations are high. This expansion results in epinasty, the downward growth of the leaves such that they appear to droop. Unlike wilting, epinasty does not involve loss of turgor. In some species (*e.g.*, pea and tomato), flooding induces stomatal closure apparently without detectable changes in leaf water potential. Oxygen shortage in roots, like water deficit or high concentrations of salts, can stimulate abscisic acid (ABA) production and movement of ABA to leaves. However, stomatal closure under these conditions can be attributed mostly to the additional production of ABA by the older, lower leaves. These leaves do wilt, and they export their ABA to the younger turgid leaves, leading to stomatal closure.

Submerged Organs can Acquire O_2 through Specialized Structures

In contrast to flooding-sensitive and flooding-tolerant species, wetland vegetation is well adapted to grow for extended periods in water-saturated soil. Even when shoots are partly submerged, they grow vigorously and show no signs of stress. In some wetland species, such as the water lily (*Nymphoides peltata*), submergence traps endogenous ethylene, and the hormone stimulates cell elongation of the petiole, extending it quickly to the water surface so that the leaf is able to reach the air. Internodes of deep-water (floating) rice respond similarly to trapped ethylene, so the leaves extend above the water surface despite increases in water depth. In the case of pondweed (*Potamogeton pectinatus*), an aquatic monocot, stem elongation is insensitive to ethylene; instead elongation is promoted even under anaerobic conditions by acidification of the surrounding water caused by the accumulation of respiratory CO_2. In most wetland plants, and in many plants that acclimate well to wet conditions, the stem and roots develop longitudinally interconnected, gas-filled channels that provide a low-resistance pathway for movement of oxygen and other gases. The gases (air) enter through stomata, orthrough lenticels on woody stems and roots, and travel by molecular diffusion, or by convection driven by small pressure gradients. In many wetland plants, exemplified by rice, cells are separated by prominent, gas-filled spaces, which form a tissue called aerenchyma, that develop in the roots independently of environmental stimuli. In a few nonwetland plants, however, including both monocots and dicots, oxygen deficiency induces the formation of aerenchyma in the stem base and newly developing roots. In the root tip of maize, hypoxia stimulates the activity of ACC synthase and ACC oxidase, thus causing ACC and ethylene to be produced faster. The ethylene leads to the death and disintegration of cells in the root cortex. The spaces these cells formerly occupied provide the gas-filled voids that facilitate movement of O_2. Ethylene-signaled cell death is highly selective; cells not destined to die in the root are unaffected. Arise in cytosolic Ca^{2+} concentration is thought to be part of the ethylene signal transduction pathway leading to cell death. Chemicals that elevate cytosolic Ca^{2+} concentration promote cell death under noninducing conditions; conversely, chemicals that lower cytosolic Ca^{2+} concentration block cell death in hypoxic roots that would normally form aerenchyma. Ethylene- dependent cell death in response to hypoxia is an example of *programmed cell death.* Some plants (or parts of them) can tolerate exposure to strictly anaerobic conditions for an extended period (weeks or months) before developing aerenchyma. These include the embryo and coleoptile of rice and of *Echinochloa crusgalli* var. *oryzicola* (rice grass), and rhizomes (underground horizontal stems) of *Schoenoplectus lacustris* (giant bulrush), *Scirpus maritimus* (salt marsh bulrush), and *Typha angustifolia* (narrow-leafed cattail). These rhizomes can survive for several months and expand their leaves in an anaerobic atmosphere. As the root extends into oxygen-deficient soil, the continuous formation of aerenchyma just behind the tip allows oxygen movement within the root to supply the apical zone. In roots of rice and other typical wetland plants, structural barriers composed of suberized and lignified cells prevent O_2 diffusion outward to the soil. The O_2 thus retained supplies the apical meristem and allows growth to proceed 50 cm or more into anaerobic soil. In contrast, roots of nonwetland species, such as maize, leak O_2, failing to conserve it to the same extent. Thus, in the root apex of these plants, internal O_2 becomes insufficient for aerobic respiration, and this lack of O_2 severely limits the depth to which such roots can extend into anaerobic soil.

Acclimation to O_2 Deficit Involves Synthesis of Anaerobic Stress Proteins

Most of the anaerobic stress proteins have been identified as enzymes of the glycolytic and fermentation pathways. The mechanism for sensing reduced oxygen levels under hypoxic or anoxic conditions is not completely clear. However, one of the earliest events to occur following lowering of O_2 levels is an elevation of the intracellular Ca^{2+}. Evidence suggests that this calcium signal is involved in the signal transduction of anoxia. Within minutes of the onset of anoxia, a rise in cytosolic Ca^{2+} concentration acts as a signal leading to increases in mRNA levels of alcohol dehydrogenase (ADH) and sucrose synthase in maize cells in culture. Chemicals that block a rise in intracellular Ca^{2+} concentration also prevent the expression of the genes for ADH and sucrose synthase from being induced by anoxia, and they greatly enhance the sensitivity of maize seedlings to anoxia (Sachs *et al.*, 1996). The accumulation of mRNAs of the anaerobic stress genes results from changes in the rate of transcription of these genes. Analysis of common sequence elements in the promoters of the ADH genes of maize and *Arabidopsis* and of the other anaerobic stress genes has led to the identification of an anaerobic stress element and a G-box element that bind *cis*-acting transcription factors leading to the transcriptional activation of these genes. Note also that there is strong evidence that some type of translational control of anaerobic stress genes is also occurring. The efficiency with which mRNAs for non-anaerobic stress–regulated genes are translated following hypoxic stress is dramatically lower than that of stress-regulated genes such as ADH.

References

Anderson, J.M. (1995). The grand design of photosynthesis: Acclimation of the photosynthetic apparatus to environmental cues. *Photosynthesis Research,* 46: 129-139.

Buchanan, B.B., Gruissem, W. and Jones, R.L. (2000). *Biochemistry and Molecular Biology of Plants*, Rockville, MD: American Society of Plant Physiologists, 26: 100-103.

Ingram, J. and Bartels. D. (1996). The molecular basis of dehydration tolerance in plants. *Annual Review of Plant Physiology and Plant Molecular Biology,* 67: 145-150.

Mittler, R. (2006). Abiotic stress, the field and environment and stress combination. *Trends in Plant Science,* 46: 120-123.

Glossary

Antenna: Large numbers of light collecting pigment serve as antenna, where the excitation transfer process in the antenna is a purely physical phenomenon and does not involve an chemical change. Chemical reactions first take place in the reaction center of PSII and PSI; subsequent reactions stabilize the unstable products of the initial chemical reactions.

Ammonification: Decomposition of amino acids and other nitrogen containing organic compounds, resulting in the production of ammonia (NH_3) and ammonium ions ($NH4^+$) ready to be taken up again.

Abscission: The dropping off of leaves, flowers, fruits, or other plant parts, usually following the formation of an abscission zone.

A zone: The area at the base of a leaf, flower, fruit or other plant part containing tissues that play a role in the separation of a plant part from the main plant body.

ATP (adenosine triphosphate): A nucleotide consisting of adenine, ribose sugar, and three phosphate groups; the major source of usable chemical energy in metabolism. On hydrolysis, ATP loses one phosphate to become adenosine diphosphate (ADP), releasing usable energy.

ATP Synthase: An enzyme complex that forms ATP from ADP and phosphate during oxidative phosphorylation in the inner mitochondrial membrane. During photosynthesis formed in the PS I photo-reaction: ADP + Pi $\rightarrow$ATP

Allelophathy: (Gk. allelon, of each + pathos, suffering) The inhibition of one species of plant by chemicals produced of another plant.

Bacterium: An auto- or hetero-trophic prokaryotic organism.

Bacteroids: A form of specialized itrogen fixing cell.

Beneficial element: Nutriemts that are beneficial to the growth and development of some plants,but have not been shown to be requirements of all plants.

Bioenergetics: The application of thermodynamic laws to the study of energy transformations in living organisms.

Biological clock: The internal oscillator that regulates innate biological rhythms.

Bological stress: Aiological insult (*e.g.* insects, disease) to which a plant may be exposed.

Bolting A rapid elongation of the stem, especially i plants with rosette habit. Boltng usually preceds flowering.

Brassiniosteroids A class of steroidal plant hormones.

Cyanobacterium: Autotrophic organism capable of fixing nitrogen from air (heterocyst) and utilizing light energy to accomplish its energetical requirements.

Chloroplast: The thylakoids within the chloroplasts of cyanobateria are not stacked together in grana, but randomly distributed (lack PS II, cyclic photo-phosphorylation).

Oxygenic photosynthetic reaction: $CO_2 + 2H_2O \rightarrow CH_2OH{\approx}P \rightarrow (CH_2O)n + H_2O$

CO_2 Pathways: The conversion of CO_2 to basic sugars during photo- or chemosynthesis can be fixed in plants in various ways based on their carbon-fixing metabolism; see plants, types of.

Carnivorous Plants: A semi-autotrophic plant, feeding upon animals;

Chemosynthesis: Chemosynthetic organisms derive their energy requirements from the oxidation of inorganic material; as performed by hydrogen bacteria etc. $6CO_2 + 12H_2S \rightarrow C_6H_{12}O_6 + 6H_2O + 12S$

Chlorophyll: photosynthetic pigments.

Chloroplast: A two-membranous plastid containing pigments (chlorophyll and carotenoids); see photosynthesis.

Circadian Rhythm: The biological clock.

Coenzymes: An organic molecule, or nonprotein organic cofactor, that plays an accessory role in enzyme-catalyzed processes, often by acting as a donor or acceptor of electrons.

Cycle: Bio-geochemical cycles on a global and local scale.

Carbon C.: Is in equilibrium between the photo autotrophic plant kingdom, the heterotrophic animal kingdom and microorganisms. The final breakdown (mineralization) of organic matter bound in living systems is executed by bacteria and fungi by aerobic breakdown, and frequency by anaerobic degradation (fermentation). Besides the CO_2 which returns to the atmosphere, there is also carbon which is deposited in the form of carbonates in the oceans.

Cytoplasmic streaming: The plasma in the cell rotates and circulates actively under the influence of light. Particles in the protoplast such as nucleus, mitochondria, and plastids are often carried passively.

Chloroplastic Orientation: Active repositioning of chloroplasts in the cytoplast of the mesophyll cells.

Diastrophic: Dark-light position, with max. light absorption *i.e*, perpendicular to incoming light.

Parastrophic: Bright-light position; reduced light absorption; *i.e*,almost parallel to incoming light.

Cyclic Photo-Phosphorylation: A photosystem lacking PSII, yielding only small amounts of ATP an no NADPH; a very ancient type of photosynthesis.

Cytokinins: *Promote growth*; Stimulates cell division; kindle growth in lateral buds, and block leaf senescence. Chemically related to adenine. Transported from roots upward in vascular system.

Chloroplast: A two-membranous plastid containing pigments (chlorophyll and carotenoids); most active photosynthetic tissue in higher plants is found in the pallisade parenchyma of leaves (mesophyll).

Carotenoids: A class of fat-soluble pigments (yellow and orange pigments); found in chloro- and chromoplasts of plants. Carotenoids act as accessory pigments in photosynthesis

Chlorophyll: (Gk. chloros, green + phyllon, leaf) The green pigment of plant cells, which is the receptor of light energy in photosynthesis; a tetrapyrrole ring structure on top with 4 internally placed N-atom, itself facing towards the centrally located Mg-atom; the entire complex is attached to a hydrophobic C20H39 phytol tail, which anchors the molecule into the photosynthetic thylakoid membrane; see table and scan below.

a: blue-green; with an extra CH_3 molecule attached at the opposite end of the tetrapyrrole.

b: yellow-green; aldehyde (CHO) instead of CH_3 attached at the opposite end of the tetrapyrrole.

Cytochrome: (cyt b6-f) Four different integral proteins, three of which contain iron that undergoes reduction to Fe^{2+} and then oxidation back to Fe^{3+} during electron flow. It mediates electrons generated by PSII towards the PSI reaction center.

Cyclic Electron Flow: In chloroplasts, the light induced flow of electrons originating from and returning to photosystem I.

C_3P.: (Calvin cycle) Enzymatically mediated photosynthetic reactions of shade-plants during which CO_2 is attached to ribulose, a C5-sugar (RuBP, a CO2 acceptor), yielding a C6-sugar which spontaneously breaks into 2 C3-sugars (3-PGA). Rubisco is resynthesized out of 5C3-sugars giving 3C5-sugars. Sugar-compounds can temporarily be converted to starch, stored in amyloplasts, and reconverted into sugars via ATP and NADPH to ADP and $NADP^+$ (glucose is more stable and compact than ATP, NADPH) - see below:

$$6CO_2 + 12NADPH + 12H^+ + 18ATP \rightarrow 1glucose + 12NADP^+ + 18ADP + 18Pi + 6H_2O$$

CAM P: (Crassulacean Acid Metabolism) Variant of the C_4 pathway; phosphoenolpyruvate fixes CO_2 in C_4 compounds (PEP carboxylase) at night.

The malic acid so formed is stored in the vacuole. During daytime, fixed CO_2 (malic acid) will be decarboxylated and transferred to the ribulose biphosphate (RuBP) of the calvin cycle within the same cell. Characteristic of most succulent, slow-growing, desert-plants; *e.g.* cacti.

C_4P.: Sun-loving plants; CO_2 is fixed to a compound known as PEP (by the enzyme PEP carboxylase in mesophyll cells) to oxaloacetate (a C_4 compound), which is rapidly converted to malic acid. This malate is then transported to bundle sheath cells (spatial separation), where the CO_2 is released (turbo charger) converting back to pyruvate. The CO_2 thus enters the calvin cycle, ultimately yielding sugars and starch. Pyruvate returns to the mesophyll cells for regeneration of phosphoenolpyruvate (PEP); requires more energy than in C_3 plants.

Diaheliotropism: the movement of the leaves is such that the broad surfaces of the blades remain perpendicular to the sun's direct rays, resulting in a higher photosynthetic rate.

Denitrification: Conversion of nitrate to gaseous nitrogen in which ammonia is oxidized, summarized in four enzymatic reactions yielding nitrogen; carried out by a few genera of free-living soil bacteria. NO_3 - (*nitrate*)→ NO_2 - (*nitrite*) → NO (*nitric oxide*) → N_2O (*dinitric oxide*) →N_2 (*dinitrogen*)

Defense Mechanism of plants: Effective protection is often achieved by physical means (thorns,. Leaf hairs, sticky sap) but also chemically (cyanides, camphor, tannins, cocaine, caffeine, nicotine, morphine, salycilic acid, etc.). Injured areas are usually walled off (compartmentalization) with resins.

DELLA protein: The reversal of etiolation caused by light.

Designed leakage: A hypothesis that attempts to explain the role of cavitation in time of water stress.

Dicotyledons (dicots): One of two classes of flowering plants,characterized by having seeds with two cotyledons,net-veined leaves,and flower parts in fours or fives.

Dinitrogenase:The enzyme responsible for fixing nitrogen.

Day Neutral Plants: Flower without respect to length of the day.

Ecodormancy: Dormancy imposed by limitations in the environment.

Electron transport chain: The third stage of cellular respiration in which electrons derived from the oxidation of intermediates in the citric acid cycle are transferred to oxygen with the accompanying conversion of redox energy to ATP. The electron transport chain is based in inner mitochondrial membrane.

Elicitor: Any metabolite isolated from pathogens that evoke a hypersensitive response in hist plant.

Embolism: An obstruction caused by the formation of large gas bubbles in the xylem.

Endodormancy: Dormancy that is an inherent property of the dormant structure itself.

Endogenous Rhythm: A rhythm that is not imposed by external factors but is generated within an organ or organism and persists under constant conditions (usually constant light or constant darkness).

3-Enolpyruvylshikimate-5-phosphate synthase (EPSPS): A key enzyme in the synthesis of aromatic amino acids. EPSPE is inhibited by the herbicide glyphosate.

Enthalpy:The total heat energy of a system,including energy available to do work.

Entropy: A measure of the randomness or disorder of a system associated with the ranom movement of matter,or a measure of energy that is not available to do work.

Epinasty: The downward curvature of leaves or other organs,especially in response to the hormone ethylene.

Essential nutrient element: Nutrient elements that are required by the plants i order to successfully complete their life cycle.

Feedback inhibition: In biochemistry, feedback inhibition is a mechanism to control metabolic pathways where the end product inhibits the activity of an enzyme earlier the pathway.A similar situation arises in genetics when the product of one gene represses the transcription of another gene.

Field capacity: The water that reains in capillary pores of soil after free (gravity) drainage is completed.

Florigen: A hypothetical flower-promoting hormone.

Flower: The reproductive structure of an angiosperm a complete flower includes sepals, petals,stamens and carpels,but all flowers contain at leastone stamen or carpel.

Fluence: The quantity of radiant energy falling on a small sphere.divided by the cross-section of the sphere.

Flux: The amount of material crossing a unit area per unit time.

Free energy: Energy that is available to do work.

Fruit set: The first stage in fruit development, defined by the decision to either abort further development or to proceed with further cell division and cell enlargement in the ovary walls.

Fucoxanthin: (Gk. phykos, seaweed + xanthos, ye/br) Brownish pigment of brown algae and Chrysophytes.

Gibberellines: *Promote growth*; stimulate both cell division and cell elongation, causes bolting (without the need to expose plant to cold or long days), new leaves, branches, flowering, larger and looser fruits, among other effects. Seeds need higher levels to germinate (conversion of starch to sugars); low levels in young plants cause dwarfism, high levels the opposite; is transported up-/ downward in vascular system.Auxin and gibberelline stimulate plant growth by increasing the extensibility of cell walls.

Gene for gene model: A model that attempts to explain the genetic basis for the hypersensitive reaction in plants.

Gluconeogenesis: The biosynthesis of glucose from noncarbohydrate precursors.

Glycolysis:The first stage of respiratory carbon metabolism in which hexose sugars are partially oxidized to pyruvic acid. Glycolysis takes place in the cytosol.

Glyoxysome: An organelle that serves as a site for the conversion of fatty acids to carbohydrate.

Grana: Regions in the chloroplast where adjacent thylakoids appear to be closely appressed.

Graviropism: A differential growth response to gravity.

Gross primary productivity: (GFP) The total carbon assimilated by photosynthesis.

Ground state The energy state of an unexcited molecules.

Growth retardant: A chemical that causes dwarfing by interfering with gibberllin biosynthesis.

Gravitropism: (L. gravis, heavy; also geotropism) Response of a shoot or root to the pull of the earth's gravity. Roots grow downwards, positive gravitropism (high levels of auxin inhibit growth in roots), whereas shoot grows upwards, negative gravitropism. Roots are more sensitive to the response of auxin than stems (high auxin concentration, inhibits root growth). Perception of gravity is correlated with sedimentation of amyloplasts (starch containing statolithic plastids) w/n specific cells of shoot and root.

Growth of Plants: Growth proceeds in an irreversible sigmoidal increase in size (slow, fast, slow, adult plant). The roots are the first structure to emerge (rhizoid pole) anchoring the germinating seed. Light controls elongation of hypocothyl (HC), the thallus pole; see also photosynthetic pigments - phytochrome:Germinated, darkness promotes a huge increase in length of the HC, otherwise remain dormant.Blue light triggers germination, orientation toward light source, lateral branching and greening of HC.Red-light causes the HC to remain short (slow vertical cell elongation) but green (photosynthetic active).

Guard Cell: Pairs of specialized epidermal cells surrounding a pore, or stoma; changes in turgor pressure of a pair of guard cells cause opening and closing of the pore

Heterocyst: Site of N_2 fixation; a specially differentiated cells, working under anoxic onditions (H_2 would combine with O_2 to form water). Ferrodoxin taking part in the nitrogenase reaction, is reduced via a strongly exergonic reaction which is fueled by energy from photosynthesis of neighboring cells, probably maltose, which is broken down in the heterocyst (see also cycle N_2-fixation): N_2 + 8e- + $10H^+$ →(nitrogenase) →H_2 + $2NH_4{}^+$(strongly exergonic, ammonium, used in amino acids).

Halobacterium: A color-sensing chemo-autotrophic bacterium (contains rhodopsin) capable of phototaxis (move to/fro a light gradient with a H+ powered flagellar motor); the purple pigment changes the absorptive spectrum under the influence of light 570 nm to 412 nm) yielding H+ protons. The externally generated Ph gradient drives phosphorylation of ADP to ATP via passive back diffusion. C-fixation is achieved by using H2S; since this process results in a solid product, sulfur will be deposited in the soil. Anoxygic photosynthesis does not require two photosystems since the energy level to split H_2S is lower than that one of water (H_2O).

Anoxygic photosynthesic reaction: $CO_2 + 2H_2S \rightarrow (Elight = hf) \rightarrow (CH_2O) + H_2O + 2S$

Heat: Many plants, (*e.g.*: aroids) when inflorescence is ready for pollination, generate heat by oxidizing large amounts of stored food, manly fats (up to 1/4 of their total weight a day). The heat (up to 30°C above ambient temp.) causes bad smelling amines to emanate attracting pollinating insects (flies, etc.). A common fuel is known to be salicylic acid (similar to aspirin). Since many of these plants flower in late fall, trapped pollinating insects are kept warm during the night to be released, loaded with pollen, the morning after.

Hormones of Plants: (Gk. hormaein, to excite) Chemical-organic substances that regulate growth, flowering, etc. released by the action of light (phytochromes), water, temperature, or other influences. Some hormones act locally while others are transported to distant tissues, where they produce specific physiological responses; others act within the same tissue where they are produced.

Heliotropism: (Gk. helios, the sun) Ability of the leaves and flowers of many plants to move diurnally, orienting themselves either perpendicular or parallel to the sun's direct rays; also called heliotropism.

Hyponasty: An abnormal bending up of organs such as leaves and petioles.

Hypotonic solution: A solution with a lower solutr content than a cell or aother solution and hence, less negative osmotic potential.

Inducible protein: A protein that is present only when the appropriate substrate is avalible.

Ion pumps: A term used in references to active transport systems because they move solute against a constration or electrochemical gradient.

Isoflavones: A group of flavones with antimicrobial properties.

Insectivory: Feeding upon animals; plants that are able to utilize proteins obtained from trapped animals, chiefly insects; *e.g.* venus fly trap, in which inward-lying cells are elongated by virtue of the high vacuoleturgor- pressure; once an insect touches the trigger-hairs the flaps snap shut rapidly (deflation of stretched cells consumes ATP), pressing the insect against digestive glands on the inner surface of the trap. The trapping mechanism is so specialized that it can distinguish between living prey and inanimate objects.

Jasmonic acid: A molecule that,along with its methyl ester,appears to have an important role in plant defense against insects and disease.

Law of the Minimum: The growth of a plant is dependent upon the amount of "foodstuff" (trace elements) presented to it in minimum quantities (J. Liebig).

Light Compensation Point: CO_2 fixation balanced by mitochondrial respiration; see photosynthetic response.

Light Dependent Reaction: Photosynthetic reaction in the photosystems I & II that require light and cannot occur in dark; see photosynthetic proteins and photosynthesis.

Light Independent Reaction: The enzymatic reaction (calvin cycle) in photosynthetic cells concerned with the synthesis of glucose (more stable and compact than ATP and NADPH) from CO_2, ATP, and NADPH; see plants- types of and photosynthesis.

Light dependent reaction: Light is required to generate ATP and NADPH.

PSII: $2H_2O \rightarrow (Elight = hf) \rightarrow O_2 + 4H^+ + 4e-$

Light independent reaction (dark reaction in the stromaof photosynthetic cells), the enzymatic reactions resulting in the synthesis of glucose from CO_2, ATP; and NADPH;

$ADP + Pi \rightarrow (Elight = hf) \rightarrow ATP$

PSI: $2NADP^+ + 2H^+ \rightarrow (Elight = hf) \rightarrow 2NADPH$

$CO_2 + H_2O \rightarrow CH_2OH''P \rightarrow (CH_2O)n$

$2NADPH \rightarrow 2CH_2OH''P \rightarrow 2NADP^+ + 2H^+$

Nitrate (NO_2 -) conversion to ammonium (NH_4^+) occurs in chloroplasts as well, hence, in stark competition for ATP and free electrons with the carbon fixing process! Allover oxygenic photosynthesic reaction: $CO_2 + 12H_2O \rightarrow (Elight = hf) \rightarrow C_6H_{12}O_6 + 6H_2O + 6O_2$

An/oxic autotrophic organisms use light energy to satisfy their energetical requirements; anoxic bacteria derive their H^+ protons from chemical compounds other than H_2O requiring less energy to be split); see bacterium. Phototrophic anoxic reaction: $CO_2 + H_2S \rightarrow (Elight = hf) \rightarrow CH_2O + H_2O + S$

Light Shielding: Absorption (conversion into heat) of excessive light; see tropism (cytoplasmic streaming) and photosynthetic pigments (xanthophyllic carotenoids).

Light Compensation Point: At this point, the amount CO_2 evolved by mitochondrial respiration is balanced by the amount of CO_2 fixed by photosynthesis.

Malate (malic acid): A C_4 compound and the end-product of the C_4-pathway. Used as a counter ion for K^+ uptake, and to supply H^+ for the proton pump (intercellular pH regulation) and is the product of the C_4 pathway of the CO_2 fixation; see there. During stomatal opening, starch in the guard cells is broken down via glycolysis to phosphoenolpyruvate, which is used with CO_2 uptake to form malate, therefore the level of malate increases;

Mesophyll: The ground tissue (parenchyma) of a leaf, located between the layers of epidermis; it is structured into pallisade- and spongy parenchyma, housing chloroplasts which contain chlorophyll (site of photosynthesis).

Motility of Plants: Even though plants cannot move actively across a given area (except adventitious root development) it possesses various ways of motion, which requires ATP:

NAD^+ (nicotinamide adenine dinucleotide): A coenzyme that functions as an electron acceptor in many of the oxidation reactions of respiration.

$NADP^+$ (nicotinamide adenine dinucleotide phosphate): A coenzyme that functions as an electron acceptor in many reduction reactions of biosynthesis; similar in structure to NAD^+ except that it contains an extra phosphate group. During photosynthesis in the PS II photoreaction, $NADP^+$ is converted to NADPH:$4NADP^+ + 4H^+ \rightarrow$ (Elight = h"f) $\rightarrow$4NADPH (is 7 times more efficient as an energy carrier than ATP)

Noncyclic Electron Flow: The light-induced flow of electrons from water to $NADP^+$ in oxygen-evolving photosynthesis; it involves both photosystem I and II.

Noncyclic Photo-Phosphorylation: The modern type of plant having both PS I & II - the zigzag scheme providing both ATP, and the more efficient energy carrier NADPH.

Nitrogen C.: Organic (NH_3^-) and inorganic (NO_3^-) linked via denitrification and atmospheric N_2 fixation.

Nitrate Fixation: The conversion of nitrate (NO3 -) into amino nitrogen by heterotrophoc organs (roots) or by symbiotic organisms (*Nostoc sp.*) and takes place in leucoplasts. Although this process is lightindependent, ATP used in these steps are generated during light-dependent carbon fixation:

NO_3 -(cytoplasma) NADPH + $H^+ \rightarrow$(nitrogen reductase) $\rightarrow NO_2^-$ + NADP + H_2O

NO_2 -(chloroplast) + 6ferredoxin (e-donor) $8H^+ \rightarrow$(nitrite reductase) $\rightarrow NH_4$ + $2H_2O$

NH_4 + 2-oxoglutarate $2H^+$ ATP* $\rightarrow$ glutamate +H_2O +H + ADP + Pi *) Conversion of the ammonium ion (NH4 $^+$) into organic compounds happens in chloroplasts (in tough competition with CO_2 reduction for ATP and free "e-") predominately through a 2-step reaction in which glutamine serves as intermediary product. The successor glautamate is then used as an amino group donor for various 2-oxo acids, *e.g.* oxaloacetate, within and outside the chloroplast (transamination).

Oleosomes: Storage droplets of lupid or oil (also called oil bodies,or spherosomes) that are normally found in cells of the cotyledons or endosperm of seeds.

Osmoregulation:The proess of regulating the osmotic properties of plant cells.

Osmosensor:A device that is able either to detect changes in the chemical potential of extracellular water or osmotic properties of the cell.

Osmosis: The properties of water passing through a semipermeable membrane with the tendency of eventually equlaizing the water potential in the two compartments.

Osmotic potential (ø) The change in free energy or chemical potential of water produced by solutes, carries a negative sign,also called solute potential.

Osmotic pressure The pressure developed by a solution separated from pure water by a differentially permeable membrane,it is generally an index of the solute concentration of the solution.

Oxidative phosphorylation:The cheiosmotic synthesis of ATP associated with the transfer of electrons through the electron transport chain and accompanying consumption of oxygen.

Oxygen evoling complex(OEC):A cluster of four manganese ions associated with a small complex of proteins that are responsible for the splitting of water and the consequent evolution of molecular oxygen.

Oxygen C.: It is complementary to the carbon cycle; in contrast to CO_2 in photosynthesis, O_2 is due to its high concentration in the atmosphere, no longer a globally limiting factor. In water, however, it is often not possible to supply enough O_2 for the complete mineralization of dead organisms, leading to the deposition of organic materials; respiration and photosynthesis are elementary steps within the oxygen cycle.

Oxaloacetate:): The C_4-molecule formed in C_4-plants, *i.e.*: the krebs cycle; the product when Pi is split off from PEP. Depending on the species, oxaloacetate is either reduced to malate or converted to asparate (additional amino group NH_2), both C_4-compounds. Malate or asparate is a mediator transporting CO_2 used in the calvin cycle.

Photosynthesis: Conversion of light energy to chemical energy; see there

Paraheliotropism: Avoiding direct sunlight during periods of drought by orienting leaf blades parallel to the sun's rays. Minimizes absorption of solar radiation, lowers leaf temp., and transpirative water loss.

Phototropism: (Gk. photos, light) Growth movement of cells or organs in which light plays a decisive role, and is related to the direction of light as the controlling external factor; *e.g.*: growth of a plant toward a light source by the influence of auxin; the cells of the shaded side of the tip migrates from the light-side to the dark-side, resulting in a turn or bend (compare photoperiodism).

Pathogen: (Gk. pathos, suffering + genesis, beginning) An organism that causes disease.

PEP or **Phosohophenolpyruvate**: The compound to which CO_2 binds in C_4-plants. PEP converts under the influence of CO_2 to oxaloacetate, which is either reduced to malate (malic acid) or asparate (extra amino group NH_2).

Peroxisome: A microbody that plays an important role in glycolic acid metabolism associated with photorespiration.

PGA (3-phoshoglycocerate): The C3-sugar formed in C3- plants, *i.e.*: the calvin cycle.

Phosphoryltation: (Gk. phosphorous, bringing light) A reaction in which phosphate is added to a compound; *e.g.*: the formation of ATP from ADP (PSI), NADPH from NADP+ (PSII) and inorganic phosphate; see photosystem.

Photo-P.: (Gk. photos, light) The formation of ATP in the chloroplast during photosynthesis.

Photoperiodism: Response to duration and timing of day and night; a mechanism evolved by organisms for measuring seasonal time. Plants that flower only under certain day-length conditions; and is a biological response to a change in the proportions of light and dark in a 24 hour daily cycle.

Short Day Plants: Flower in early spring or fall when days are shorter with less intense light.

Long Day Plants: Flower in the summer, when the light periods are longer than the critical length.

Photorespiration: types of - C_3-plant.

Photosynthesis: (Gk. photos, light + syn, together + tithenai, to place) Conversion of light energy to chemical energy; PLANTS: Solar radiation is used by plants to oxidize water, reduce CO_2, and release of O_2. Plants use <5% of the radiant energy, the majority is reflected or lost as heat.

Photosynthetic Pigments: A substance that absorbs light, often selectively.

P. Spectrum: The spectrum of light waves absorbed by a particular pigment eliciting a certain reaction; gamma rays (short wavelength) are too energetic (destroy biological molecules), radio waves don't excite them at all.

Phycobilins: A group of water-soluble accessory pigments, which occur in the red algae and cyanobacteria.

Phytochrome: A phycobilin-like pigment (photoreceptor for red and far-red light) found in the cytoplasm of plants and a few green algae; phytochromes do not participate in photosynthetic reactions. Plants contain phytochrome in two different inter-convertible forms:

Pr absorbs red light (660 nm), the biological inactive form of the protein (inhibits reactions, but allows plant to grow pale and spindly, changes to Pfr (pigment fully reactive) when exposed to (red) light. Phytochrome is continuously synthesized as Pr from its amino acid precursors.

Pfr absorbs far-red light (730 nm), the biological active form triggering reactions, *e.g.* induces germination flowering, dormancy, leaf formation, seed germination, shade detection, etc. by triggering the release of plant hormones via a cis-trans isomerisation (cascade amplification). Pfr is converted back to Pr when exposed to far-red light (darkness, dark reversion) or lost through hydrolysis by proteases.

Photosynthetic Proteins: The protein complexes of PSI and PSII spatially separated within the thylakoid membrane indicating a one to one stoichiometric ratio between those two systems.

Photosystem: A discrete unit of organization of chlorophyll and other pigment molecules embedded in the stroma thylakoids of chloroplasts and involved with the light-requiring reactions of photosynthesis. Oxygenic photosynthesis requires two photosystems since the energy level to split H_2O is very high. P680 and P700 refer to the wavelength of maximum absorption of the reactions center chlorophylls in PSII and PSI.

PS II: A series of noncovalently bondend complex intrinsic polypeptides; associated with three peripheral (extrinsic) polypeptides, thought to aid in binding of Ca_2^+ and Cl^-, both of which are essential for photolysis of water. The P680 core complex (LIGHT HARVESTING COMPLEX-II) receives red light energy by inductive resonance from a total of about 250 *chlorophyll a* and *b* molecules (associated with few integral chlorophylls + xanthophylls proteins which act as an antenna system), producing a strong oxidant (oxidizes water) and a weak reductant: f, frequency [s-1] [Hz]

$2H_2O \rightarrow (Elight = hf) \rightarrow O_2 + 4H^+ + 4e$- h,

$2H_2O$ + 4 photons + 2 oxidized plastoquinone (2PQ)$\rightarrow O_2 + 4H^+$ + 2 reduced plastoquinone ($2PQH_2$)$2PQH_2$ + 4 plastocyanin ($4PC\text{-}Cu_2^+$) $\rightarrow$2PQ + 4PC(Cu^+) + $4H^+$ (lumen)

The increasing H^+ concentration (low pH causes an electrochemical proton gradient) within the lumen of the thylakoid is used to synthesize ATP, when H+ tunnels back out to the stroma through the integral coupling factors (CF's): ADP + Pi $\rightarrow$(H^+) $\rightarrow$ATP

PS I: Even though it uses (far) red light independently, PSII recruits electrons originally released by the PSII H2O-lysis mediated via the cytochrome complex. This reaction causes cytochrome to transport H+ ions across the membrane from the stroma into the thylakoid membrane (further decrease of internal pH). Two large polypeptides bind the reaction center P700, some *chlorophyll a* molecules and three electron carriers (NADP+) called phylloquinone and a Fe-S group. The PSI core complex receives light by inductive resonance from about 100 *chlorophyll a* and *b* molecules formed to an other antenna system (LIGHT HARVESTING COMPLEX-I). The strong reductant produced by PSI reduces NADP+, to NADPH, which is released into the stroma: f, frequency [s-1] [Hz

2NADP+ + 2H+ $\rightarrow$ (Elight = hf) $\rightarrow$2NADPH] h, plank's constant

4PC(Cu^+) + 4Fd(Fe_3^+) $\rightarrow$ (Elight = hf) $\rightarrow$4PC(Cu^+) + 4Fe(Fd_2^+)

4Fe(Fd_2^+) + $2NADP^+$ + $2H^+$ $\rightarrow$4Fd(Fe_3^+) + 2NADPH

Photosynthetic Products: Most of the fixed carbon is converted either to sucrose, the major transport sugar, or to starch, the major storage carbohydrate in amyloplasts and in the stroma (temporarily) of plants. Sucrose (disaccharide) is favored over glucose (monosaccharide), to avoid digestion of monosaccharides during transportation down the phloem by those cells. Once arrived at target tissues, disaccarides are cleaved into glucose and fructose.

Photosynthetic Response: The dose-response of photosynthetic C-fixation as a function of photon flux. A C_4 (sun loving) plant has a higher light compensation point, a higher maximal photosynthetic rate, than a C_3 (shade) plant. Maximal saturation is determined by the slow-working carboxylase - see rubisco and plants, types of.

Photorespiration (PR): In C_3 plants only; photosynthesis in C_3 plants is always accompanied by PR, a process that consumes O_2 and releases CO_2 in the presence of light. Since no ATP is yielded by PR, it diverts some of the light-depending reactions from biosynthesis of glucose into reduction of O_2. Up to 50% of CO_2 fixed in C_3 plants may be again reoxidized to CO_2.

Quantum.The energy carried by photon.

Quisecent Zone: A zone of relatively inactive cells at the center of the root apical meristm.

Rhizobium: Flagellated, free-living heterotrophic soil bacteria. Approach chemotactically the rhizosphere of a susceptible root-hair, proliferating into it, causing curling growth. Rhizobia fuses with the plasma membrane of the root hair cell (without infecting them, separated by a peribacteroid membrane), tunnel through the cytosol of the cortex, force cortical- and pericycle cells to divide until in touch with the vascular tissues. Once there, bacterial colony stops dividing, enlarges, and differentiate into N_2-fixing endosymbiotic organelles (nodules). In exchange from fixed nitrogen (NH_4^+), the host provides C-compounds and a highly regulated watery O_2 environment. Since O_2 is a potent inhibitor of nitrogenase, "infected" cells produce large quantities of the O_2 binding *leghemoglobin* to keep its partial pressure at 21% but that of N_2 below 78%; see cycle, nitrate fixation.

Reaction center: The chlorophyll plus associated proteins and redox carrires that are directly involved in light driven redox reactions.

Reactive oxygen species: Toxic forms of oxygen such as free radicals and peroides,which can lead to cell damage.

Redox couple: A reduced/oxidised pair such as NADPH.

Redox potential: The tendency to accept or donate electrons.

Reducing sugars: Reducing sugars have a free aldehyde group that reduces mild oxidizing agents.The standard test is the reduction of silver ions to yield a metallic silver mirror lining on the inside of the reaction vessel.

Respiration: A series of metabolic reactions for retrieving the energy and carbon stored by photosynthesis.

Retrograde gene regulation.The regulation of nuclear genes by organelles such as the chloroplast.

Rhizobia: A group of nitrogen fixing bacteria.

Rieske iron-sulfur (FeS) protein: Iron-binding proteins in which the iron complexes with sulphur residues rather than a heme group as in the case of the cytochromes.

Rubisco: An acronym for the enzyme ribulose-1,-5-bisphosphate carboxylase/ oxygenase.

Scarification: The process of mechanically disrupting or removing the seed coat.

SCF complex: The multiprotein enzyme 3 (E_3) complex that mediates ubiquitination of proteins targeted for degradation by the proteasome.

Second messenger: A mobile signalling molecule that is formed or released inside the cell in response to the binding of a first messenger-*e.g.*, a hormone or other ligand- to a receptor or the activation of a receptor by an environmental signal.

Shade avoidance syndrome: A condition of increased stem elongation, reduced branching and early flowering characteristic of plants growing in shade.

Soil-plant atmosphere continuum.The integrated flow of water from the soil,through the plant,and into the atmosphere.

Spectral energy distributed (SED). A measure of light quality.

Stamen: The male organ of a flower.The site of pollen production in a flower.The site of pollen production in a flower.

Stele:The central vascular core of a root.

Stillbenes: A phynyl propanoid derivative that has antifungal properties.

Stoma: A pore in the epidermis of leaves and stems. Also used to refer to the entire stomatal apparatus which includes the surrounding guard cells.

Substrate- level phosphorylation: The direct synthesis of ATP through the transfer of a phosphate group from a substrate to ADP.

Symplast: Protoplast continuity through plasmodesmata.

Systemic acquird resistance: The development of a general immune capacity throughout the ntire plant folloing an initial invasion by a pathogen.

Terpenes: A diverse family of molecules built on the 5-carbon isoprene unit.

Thermonasty Plant movements.*e.g.*, opening and closure of petals,due to differential growth in response to alternating temperature.

Thermoperiodism: The growth response of plants to alternating high and low temperature.

Thylakoids: A highly structured internal system of membranes in the chloroplast.

Ti plasmid: The tumor -inducing plasmid of *Agrobacterium tumifaciens*.

Totipotency: The concept that cells are capable of assuming an embryonic state and developing into any tissue of a mature plant.

Trace element: Another term for micronutrients.Nutrients that are required only in trace amount.

Tranfer cells: Specialized parenchyma cell with wall in growths that surface area of plasma membrane,apparently functioning in the short distance transfer of solutes.

Translocation: The transport of photoassimilates over long distances.

Transpiration ratio(TR):The ratio the uptake of CO_2 to the loss of water by evaporation from the leaf.

Tropism: A differential growth response that is related to the direction of the stimulus.

Turgor pressure: The pressure component that arises from the force eTurgor pressure: The pressure component that arises from the force exerted outwardly against the cell walls by the expanding protoplast.

Two-component system: Sensory signalling mechanisms consisting of a transmembrane sensor protein and a cytoplasmic protein called the response regulator.

Ubiquitin: A small protein that is linked to other proteins targeted for degradation by proteosomes.

Vernalin: A hypothetical plant hormone thought to be involved in the vernalzation response.

Vernalization: The promotion of flowering brought about by subjecting the shoot apex to a period of low temperature.

Vivipary: Precocious germination,*i.e.*,while the seed is still on the mother plant.

Wall pressure: The inward pressure exerted by the cell wall against an expanding protoplast,equal but opposite to turgor pressure.

Water potential: The alebric sum of the solute potential and the pressure potential and the pressure potential of water in a cell.

Water potential gradient: as the driving force for water movement.

Xanthophyll: Xanthophylls are the typical yellow pigments of leaves. These are oxygenated carotenoids that are synthesized within the plastids.

Xanthoxin: Xanthoxin is an intermediate in the biosynthesis of the plant hormone abscisic acid.

Xylem vessel: Xylem vessels are long hollow chains of tough long deadxylem cells. Xylem tissue is the water transporter cells of plants. It carries water around a plant

Xyloglycans: Xyloglucan is a hemicellulose that occurs in the primary cell wall of all vascular plants

Yang cycle: Inhibition of 5′-methylthioadenosine metabolism in the Yang cycle alters polyamine levels, and impairs seedling growth and reproduction.

Index

D

E

F

W

X

Z

www.ingramcontent.com/pod-product-compliance
Ingram Content Group UK Ltd.
Pitfield, Milton Keynes, MK11 3LW, UK
UKHW020806290926
15470UKWH00003B/983

9 789351 309901